Setting Sail for the Universe

Setting Sail for the Universe

ASTRONOMERS AND THEIR DISCOVERIES

Donald Fernie

Rutgers University Press

New Brunswick, New Jersey, and London

Library of Congress Catalogue-in-Publication Data

Fernie, Donald.
　Setting sail for the universe : astronomers and their discoveries /
　　Donald Fernie
　　p. cm.
　Includes bibliographical references and index.
　ISBN 0-8135-3088-1 (cloth : alk paper)
　　1. Astronomy—History. I. American scientist. II. Title.

　QB15 .F39 2002
　520'.9–dc21　　　　　　　　　　　　　　　2001058687

British Cataloging-in-Publication data for this book is available from the
British Library

Manufactured in the United States of America

For Yvonne

Contents

Preface and Acknowledgments

This book is made up of twenty-eight short articles originally written for the *Marginalia* column of *American Scientist*, the magazine of Sigma Xi, The Scientific Research Society. As the name of the column suggests, the style of writing is intended to be quite chatty and informal. And although the articles were written for a science audience, the seventy-five thousand scientists and engineers that make up Sigma Xi are of every kind, whether they be butterfly experts or bridge builders, and so no assumptions regarding background knowledge have been made. Most readers, scientists or not, should have little difficulty with the material.

The articles deal almost entirely with topics in the history of astronomy, the latter having been my profession and its history a strong interest of mine. In particular, it has almost always been the human interest side of the subject that has attracted me most, rather than how some concept developed over time, and that is the emphasis here. Many of the stories involve the dangerous and difficult expeditions to which astronomers of past centuries subjected themselves in pursuit of their subject, something that has long fascinated me since it stands in such contrast to our own times. The choice of topics is random and ranges widely over astronomy-related subjects, on most of which I am not an authority or specialist but rather a general practitioner. None of the material involved original research on my part, so that all the stories rely on other sources, of which published ones are noted in the references. (Occasionally I have relied on astronomical folklore picked up in the course of a half-century career.) Also, the writing of the articles has spanned sixteen years, and in some cases there have been significant new findings since the original article appeared. In these cases the notes at the end of the book bring the story up to date.

It is a pleasure to acknowledge the many people who have contributed to the publication of this book. First my good friend Pierre Demarque of Yale University, who initially suggested I write for *American Scientist*. Michelle Press, Brian Hayes, and Rosalind Reid, successive editors-in-chief of that journal, have kindly accommodated my *Marginalia* musings over the years, while associate editors Bill Hively, Michael Szpir, and particularly Michelle Hoffman have been indispensable for their detailed editing. Lil Chappell, editorial associate at *American Scientist*, has been most helpful on matters of permission regarding the reprinting of the articles. I also thank Susan Hochgraf, Elyse Carter, Tom Dunne, and Linda Huff, who produced the original figures in chapters 4, 7, 19, and 24. I am most grateful to Dael Wolfle for permission to quote from an unpublished document concerning Chappe's expedition to Mexico. Finally, my thanks to Jack Repcheck of Princeton University Press for originally suggesting that this material might be reprinted in book form, and especially to Helen Hsu of Rutgers University Press, who, as acquiring editor, has brought the project to fruition.

Setting Sail for the Universe

The Ghost of Mount Wilson

It was with no small sadness that many an astronomer—particularly those of my generation and earlier—heard recently that the Mount Wilson Observatory in Pasadena, California, will in all likelihood close this year [1985]. I say this with feeling not only because the history of Mount Wilson has in many ways been coincident with the history of twentieth-century observational astronomy, but because I have fond memories of the place. After all, I am possibly the only person to have actually seen the ghost of the 100-inch telescope.

That happened back in 1957, when I was a graduate student. The fact that I needed some high-dispersion spectrograms for my thesis, and that my supervisor had recently been a Carnegie fellow at Mount Wilson and now needed more data of his own, resulted in our both making the trip out west to use the 100-inch.

Installed on Mount Wilson, we had a run of several nights at the coudé focus of the telescope—an arrangement of mirrors with the advantage that one does not have to move with the base of the telescope as it swings in a great arc to compensate for the earth's rotation. Now I should explain that this telescope was built in 1917, and whether it was the influence of the World War or not, the thing was built like a battleship. Not just the telescope itself, but also its housing. Everything was dark steel plates and a million rivets, and one instinctively thought in terms of bulkheads and companionways. There was much clanking and groaning and whirring of machinery, all illuminated by the dimmest possible light. Finally, the coudé focus is buried in the depths of the dome, so we had to wend our way through all of this to reach the cramped working quarters.

The first night brought my introduction to Gene, who was to be

First published in *American Scientist,* 73:281, May–June 1985.

the night assistant and telescope operator. He was a pleasant fellow, but I soon saw in his eye the old let's-try-one-on-the-new-kid look, and sure enough, as soon as we had settled down to the long routine of guiding the first exposure, he started in on the story of the ghost of the 100-inch. It involved some woman who was supposed to appear in long, flowing robes, lantern in hand, searching for something or other. I paid only scant attention. Clearly it was a put-on, aimed to take advantage of the eeriness and jumpiness that can overtake any of us in the middle of a long night, and I was on my mettle. After all, only recently I had worked at the Royal Cape Observatory in South Africa, where one set off for the domes at dusk past gloomy graves of early directors, always with the thought that the neighboring estate housed the local mental institution. It was said that one should not be surprised to have a visitor materialize out of the darkness in the dome, announcing himself as Napoleon or Chaka, king of the Zulus. So what cared I for Gene and the ghost of the 100-inch?

The long hours passed. It must have been around four or four-thirty in the morning, when weariness had stilled any conversation among the three of us and I was huddled over an entry in the log book, that some small sound behind me made me turn around. Dear God! There she was! A woman in long, flowing robes, hair similarly long and flowing, and light in hand.

I did the only thing possible, although I was only halfway up the polar axis of the telescope when my heart started beating again and I turned for another look. That revealed the flowing robes to be a pretty standard nightgown and the lantern a very modern electric flashlight. It also revealed everyone else in paroxysms of laughter.

The lady, it transpired, was Gene's wife, and her presence there in the predawn stillness had a simple explanation. This being 1957, atomic bomb tests were under way in the Nevada desert, and since these were announced ahead of time (I suppose), it had become the thing for those who lived on Mount Wilson to come up to the catwalk outside the dome of the 100-inch, from which they might see the flash of the bomb illuminate the distant early-morning sky. It was, I duly noted with satisfaction, hardly the spectacular event to be worth getting up for at such an hour.

There were, of course, hot denials of any collusion, but I always had my suspicions.

Well, I can't feel sorry about the fate of the ghost if the old observatory should come to an end. Ghosts should be used to that. But it does give one pause to think back on those titans of early-twentieth-century astronomy who measured the universe from the heights of that mountain. Men like Edwin Hubble.

Hubble, if I may be excused some extravagance of phrase, fought the battle of the universe with the 100-inch as his weapon. It was he, using photographs made with the 100-inch, who in 1925 settled the centuries-old controversy as to the nature of the spiral nebulae. In doing so he immeasurably expanded our conception of the universe. Prior to that time there had been two competing theories about the spirals. One held that they were indeed enormous galaxies like our own Milky Way, the other that they were probably smaller clusters of stars subordinate to the Milky Way (colonies of our empire, as Arthur Eddington—British, of course—liked to put it). The debate on this issue had reached fairly savage proportions by the early 1920s, and because of this Hubble found himself in a curiously difficult position. This came about because the principal evidence for the colony theory (which Hubble's result disproved) was the work of another Mount Wilson astronomer, Adriaan van Maanen. Arguments might go on around the world, but the fundamental and opposing evidence was compiled only office doors apart.

The situation was this. Measurements of the Doppler- or redshift in the light from the arms of the spiral nebulae had shown that the nebulae were rotating with considerable linear velocities: a representative point well out along an arm might be spinning about the nebula's nucleus at several hundred kilometers per second. This was no great surprise. Why shouldn't something that looked like a pinwheel be spinning? The surprise appeared at the next stage. It occurred to van Maanen to try to measure the corresponding *angular* velocity of a spiral—its rate of rotation—by comparing photographs taken decades apart. One could then use the linear and angular velocities to arrive at the radius of a circle—the actual radius of the spiral nebula.

The precise positions of nebulous knots in the spiral arms were measured against a framework of "background" stars on each plate, and the comparisons made. The results, first announced by van Maanen in 1913, showed significant angular rotations. So much so, in fact, that the spirals were relatively small and could not possibly be at the distances needed to make them individual galaxies. Calculation based on the measured linear and angular velocities suggested distances of only a few thousand light-years, a factor of a hundred or a thousand too small.

This was a devastating blow to those who believed in the spirals-are-galaxies theory, the evidence for which, however, was less direct. Still, they fought back with suggestions that errors might have crept into van Maanen's measurements: precise positions of fuzzy knots would be difficult to establish; who knew what distortion old photographic emulsions might undergo in the course of decades; and so forth. Stung, van Maanen selected other spirals on other photographs, and before long a steady stream of papers emerged from his office to appear in the prestigious *Astrophysical Journal*. The results were all the same. Virtually all the nebulae showed significant angular rotation, all showed the arms to be winding up, all had similar rotation rates, and all gave distances that placed them in the outer reaches of the Milky Way. The sheer consistency seemed unanswerable. How could random errors possibly have produced this? Several big-name astronomers like Harlow Shapley felt the case was closed. The Milky Way and its attendant hordes of nebulae constituted the visible universe; it was, so far as we could see, everything. There was nothing else like it.

Hubble, meanwhile, was on another track. He too studied photographs of the spirals, which he had taken himself with the 100-inch—the world's largest telescope—under the steady, transparent sky above Mount Wilson. These photographs, under the power of a microscope, revealed a remarkable phenomenon: in a few nebulae the outer arms could be resolved into myriads of minute star-images. Moreover, some of these stars were found to be varying in brightness in a way that revealed them as Cepheid variables. Now these have an important property: the length of time they take to vary is quite regular, and is correlated with their intrinsic luminosities. That correlation had been calibrated (with some irony, as it turned out) by Shapley.

Successive photographs soon established the periods of variation and thus, thanks to Shapley, the intrinsic luminosities of these stars. The Cepheids were now standard candles. Their apparent brightness, measured from the plates, and the inverse-square law then provided their distances. The answer: the spirals lay far beyond the Milky

Way, so far away that, to appear the size they did, they must be galaxies themselves. No reasonable uncertainty in the analysis changed that conclusion. The Milky Way was not unique; it was only one among untold millions of galaxies. The universe was nothing like that imagined by van Maanen.

Hubble, though, was now caught in a serious dilemma. How could he, a junior staff member, break the news to the world without making his senior colleague look foolish? And what, indeed, was the explanation of van Maanen's results, all those consistent papers, now seen to be false? For many months Hubble pondered what to do. Correspondence with a select few influential and respected American astronomers, notably Henry Norris Russell of Princeton, finally convinced him to act, and his monumental discovery was made known to the world at a meeting of the American Astronomical Society on New Year's Day, 1925. Hubble himself was unable to attend, and his paper was read by Russell.

No satisfactory explanation of van Maanen's results has ever been given. Some technical explanations involving guiding errors and optical aberrations were mooted but were far from convincing, and remeasures of the plates by others, including Hubble, failed to reproduce van Maanen's data. The uncharitable, of course, inclined to the view that the latter had been cooked and the whole thing was a hoax. But that is not convincing either; one would be a fool indeed to cook observations that were likely to be checked by others at any time.

There is, however, a small footnote to it all that perhaps sheds some light. The Mount Wilson Observatory was as much a solar observatory as a stellar one, and indeed the solar aspects were probably the first love of its founder, George Ellery Hale. In the 1930s Hale published work showing that the sun had a general magnetic field of some 30 gauss. This is now known to be a gross overestimate by more than an order of magnitude, and a surprising result for one so careful and reputable as Hale. In trying to trace how such an error came about, a Norwegian solar physicist some years ago remeasured the original magnetograms on a modern, impersonal measuring machine, and discovered that they showed no such magnetic field at all. The trace, however, also turned up the fact that Hale himself had been too busy to measure the magnetograms and had asked a colleague to do it: van Maanen, of course!

One can only surmise that van Maanen was so eager to produce positive results that would attract attention that he deluded even himself. A minor but interesting case in the psychology of scientists.

Hubble was a most unusual man himself, though in a quite different sense. Some years after the denouement of the spiral nebula controversy, he became even more famous—particularly among the public—when his continued 100-inch work on galaxies revealed that the universe is expanding. As a result he was, I believe, the first astronomer to appear on the cover of *Time* magazine. What more can one say?

But long before that he had established a unique record. He had taken his baccalaureate in mathematics and astronomy at the University of Chicago, where, amidst a brilliant academic career that led to a Rhodes scholarship, he also excelled in athletics. His prowess extended to several sports but particularly boxing, so that when he graduated he was faced with a truly remarkable choice: he could accept his Rhodes scholarship and go to Oxford, or he could accept the offer of a Chicago boxing promoter to train as a challenger for the world heavyweight championship. Oxford won out, although while there Hubble fought an exhibition match with the world light-heavyweight champion, Georges Carpentier of France, and won his blue in a number of other sports. His degree at Oxford, incidentally, was in jurisprudence, and he briefly practiced as a lawyer in Louisville, Kentucky, before returning to the University of Chicago for graduate work in astronomy.

Hubble's subsequent fame, particularly in the 1930s, brought out another of his great attributes. Hollywood promoters thought it the thing to parade their rising stars alongside this most famous of scientists, the discoverer of the expanding universe, while press cameras blazed away. The ploy unfortunately backfired for many a male star, however, when it was found that Hubble was the more handsome of the two. [See the Notes for remarks relating to some of these stories.]

Sadly, and perhaps unexpectedly in the light of his athletic career (he continued in noncompetitive sports), Hubble died in 1953 of a heart attack at the age of sixty-three. I find it a most happy choice that the coming space telescope, itself no doubt the herald of a new age in astronomy, should have been named the Hubble Telescope. Meanwhile, the old 100-inch remains, mute testimony to his great achievements.

Mount Wilson, for so long the world's premier big-telescope observatory, included many a famous and colorful character among its staff, but rather than recount further anecdotes, let me turn to the serious question of whether it deserves its demise.

The arguments for closing it are straightforward enough. It costs perhaps one or two million dollars a year to operate such a facility, and if its equipment is now old and creaky and its skies polluted with the lights of so many cities below, surely it makes sense to close it and put the money into a modern facility at a dark site.

This is true, at least to a point (for I understand that the equipment is still entirely usable and the light pollution not as bad as one might think). Nevertheless, closing Mount Wilson will close an important door on modern astronomy.

I say this because astronomical research, at least in the observational area, has taken a curious bias over the last decade or two. The development of rapid and widespread air travel has allowed the establishment of observatories at remote and previously impractical sites. The increasingly difficult search for dark, unpolluted skies has favored, almost by definition, the choice of high mountain peaks or desert locations as far away from populated areas as possible. This has been good for astronomy. The telescopes placed in these new sites, technologically sleek and powerful, have made immensely important discoveries that could not otherwise have been made from earth.

I say this with enthusiasm; yet an attendant drawback has developed. Such stations, so expensive to build and maintain, are inevitably few in number and in heavy demand. This means that an observer, already a lucky winner over other eager applicants, will be assigned only a few nights a year at such a telescope. In any case, few astronomers can afford the time (or the strain on family relations) to spend extended periods on high mountains or in foreign deserts. Moreover, one's luck at getting another such assignment will depend in no small degree on how productive one's past assignments have been. Publish or perish, with a vengeance.

The upshot is clear. Only those research problems that will yield useful results in compressed observing runs are likely to be undertaken. Projects that involve objects varying on time scales of weeks or months, or that require the patient accumulation of large data banks for their solution, will likely go to the wall. We are in danger of missing valuable discoveries that simply happen to take a long time to emerge.

This is where Mount Wilson and other observatories like it play an important role. They are a much-needed complement to the necessarily crowded facilities at distant supersites. They offer less rushed observing, where the long-term, patient accumulation of data by astronomers living comfortably nearby can lead to important discoveries. The slow, steady cataloging of nebulae by the Herschels was the first step toward Hubble's resolution of the controversy surrounding their nature; his own patient studies of many galaxies led finally to his discovery that they recede from us at speeds proportional to their distances—the universe expands. Without others having provided the basic observations of many stars, the Hertzsprung-Russell diagram, central to almost all of stellar astronomy, would not have been devised. The Infrared Astronomical Satellite was put aloft at immense cost just to conduct a survey—and one could give many more examples. Scientists agree that important discoveries often occur unexpectedly; there is nothing that says they never happen as a result of long-term observations.

The Mount Wilson Observatory in particular deserves to continue in operation, because its skies and atmospheric conditions are not much worse than they ever were, which is to say superb. Projects can proceed at relative speed there compared with cloudier sites, although some of this work, involving slowly variable celestial objects, can even tolerate a fair amount of cloudy weather. There is much to be said for having reliable equipment right to hand in one's backyard, so to speak. No great travel costs or time is wasted by an occasional cloudy night.

Enough said. We hear that efforts are being made to form a consortium to take over the Mount Wilson Observatory so that its usefulness may not lapse. Many of us have a strong desire to see those efforts succeed. Astronomy needs Mount Wilson, no less than does its resident ghost.

Candid Posterity and the Englishman. I

Wherefore if according to what we have already said [the comet] should return again about the year 1758, candid posterity will not refuse to acknowledge that this was first discovered by an Englishman.
—*Edmond Halley*

Those of us who toil to educate the young in elementary astronomy know there is one thing at least as certain as death and taxes: no amount of cajoling or dire threat will stop a student from mispronouncing the name of Edmond Halley as "Hailey." I once halted such a miscreant in midspeech; he listened to my reprimand without interest, and laconically continued, "Yeah, well, like I was saying, there was this dude Hailey. . . ." And now, of course, as the famous comet once again approaches, the airwaves are becoming filled with news of Hailey T-shirts and other souvenirs.

It seems that candid posterity has a rock group of the 1950s, Bill Haley and his Comets, to thank for this, but those of us who had for so long demanded "Halley" of our students received a rude shock some years ago when recently discovered contemporary correspondence addressed Halley as "Hawley," suggesting that the man himself probably syllabified his name as "Hall-ey." Doubtless today's electronic media will save continuing posterity, ever more candid, from such pitfalls.

In a way, the fact that relatively so few people know how to pronounce his name, and beyond his name will likely only hazard a guess that he discovered a comet, typifies the neglect that posterity has in fact accorded this Englishman. This is all the more remarkable when one finds that even in an age renowned for unusual savants, Halley

First published in *American Scientist*, 73:471, September–October 1985.

stands out as a man of extraordinary breadth and depth. For his science alone he was hailed by the Victorians as "the second most illustrious of Anglo-Saxon philosophers," but his catholicity of interests went far beyond science: there was Halley, sea captain, voyager to the Antarctic seas, apprehended as a pirate, putting down a mutiny; Halley, consort of emperors; Halley, classicist; Halley, adviser on fortifications; Halley, putative spy; Halley, company founder and director; Halley, author of the first actuarial mortality tables; and so on. His misfortune, as Sir Robert Ball put it, was that "he had to shine in the same sky as that which was illuminated by the unparalleled genius of Newton." Like Brahms looking to Beethoven, he heard always behind him the footsteps of the giant. Yet even here, posterity might take note, had Halley not pushed and harried Newton, there might well have been no *Principia*. And it was Halley who paid for its publication.

Halley came of a fairly well-to-do family, his father being a prosperous soap-boiler in London, although his financial situation in later life is less certain. The father was found drowned in a river ("The Coroner sat upon him, & the Inquest brought him in Murthere'd") while Halley was still in his twenties, and difficulties with his stepmother left him in more straitened circumstances. Later he was pleased to have the paying job of a Royal Society assistant secretary, although even this was not without hazard. The Royal Society, itself occasionally of uncertain means, finding a glut of unsold copies of Francis Willughby's *Historia Piscium* on its hands, resolved on 6 July 1687, "that Halley be given 50 copies of *History of Fishes* instead of £50 salary."

In light of the comet for which Halley is now best remembered, it was fitting that his first scientific paper, published in 1676 when he was a twenty-year-old at Oxford, was on the determination of the elements of planetary orbits. (The elements of an orbit are the half-dozen parameters that describe the orbit's size, shape, orientation, and position of the planet in it.) But it was not until a quarter-century later that Halley began a comprehensive study of comets, published in 1705 as *Astronomiae Cometicae Synopsis*, in which he suggested that several apparitions of bright comets were really apparitions of just one comet, which would likely return again about the year 1758.

The suggestion was made on the basis of similarity of orbital elements, though it was not arrived at as straightforwardly as one might suppose from today's standpoint. First, Halley had to collect a body of observational data on which to base his calculations, and this proved no easy task. Where today any comet automatically has its position in the sky measured week by week with high precision, Halley had to contend with far more casual observations, especially as he went back through the centuries before his own time. Even obtaining contemporary observations was a problem for Halley when it came to dealing with his erstwhile friend, John Flamsteed, now astronomer royal and harboring a grudge against Halley. We find Halley writing to Newton on 28 September 1695: "I must entreat you to procure for me of Mr. Flamsteed what he has observed of the Comett of 1682 particularly in the month of September, for I am more and more confirmed that we have seen that Comett now three times since ye Yeare 1531, he will not deny it you, though I know he will me." This was indeed the object we now call Halley's comet.

Second, in Halley's day there remained uncertainty as to the mathematical description of a comet's orbit. Although the possibility of an elliptical orbit had been mooted at least as early as 1610 by Sir William Lower, a straight-line path long remained popular. After Newton's *Principia* in 1686, it had generally been accepted that the orbit must be a conic section (ellipse, parabola, or hyperbola), and observation soon rejected the hyperbola. Halley still faced the choice between ellipse and parabola, Newton favoring the latter. The choice was crucial, of course, because an elliptical orbit would bring a comet back, while a parabolic one would not. The reason this uncertainty attached to comets and not planets was that planets have elliptical orbits of very low eccentricity, whereas comets generally have elliptical orbits of such high eccentricity that the very limited portion of the orbit near the sun where we can see them can as well be represented by a parabola as by an ellipse. Halley's comet itself, for instance, has an orbital eccentricity $e = 0.97$; a parabola, by definition, has $e = 1$. For comparison, the earth's orbital eccentricity is less than 0.02.

Still, Halley was not long in coming to favor the elliptical alternative. In the same letter to Newton quoted above, Halley reports: "I find certain indication of an Elliptick Orb in that Comet [of 1680–81] and am satisfied that it will be very difficult to hitt it exactly by a Parabolick."

Proceeding with a tabulation of the elements he had calculated from all available observations of comets, Halley came to his famous conclusion: "Many considerations incline me to believe the Comet of 1531 observed by Apianus to have been the same as that described by Kepler and Longomontanus in 1607 and which I again observed when it returned in 1682. All the elements agree. . . . Whence I would venture confidently to predict its return, namely in the year 1758."

In a later edition of the *Synopsis*, Halley extended the identity of the comet to those seen in 1305, 1380, and 1456. Modern work has found recorded apparitions back to 240 B.C.E., thirty orbital cycles ago. A recent study of its physics concludes that the comet has altogether made some two thousand such orbital cycles, with about the same number still to go before its ices are eroded away to nothing, involving a time scale on the order of a hundred thousand years.

Halley in the course of his long life made many contributions, several of fundamental importance, to astronomy. They ranged from meteors (for which Halley suggested a nonterrestrial origin when a terrestrial origin was the prevailing opinion) to the universe itself (involving what is now known as Olbers's paradox—what is the implication of the sky being dark at night?). I particularly admire Halley for the fact that when he eventually succeeded the crusty Flamsteed as astronomer royal, he began an important research project that necessitated positional observations of the moon on virtually every clear night for eighteen years, starting when he was already sixty-four. Almost needless to say, the cheerful Halley brought it to a successful conclusion!

But Halley had nearly as much claim to fame as a geophysicist as he did as an astronomer. Starting when he was twenty-eight, he worked at the theory of tides, calling attention to a curious phenomenon reported from Batsha on the coast of China, where only one high tide a day was observed and where the tides ceased altogether for several days twice a month. Halley determined the systematics of the phenomenon, allowing Newton eventually to suggest that "the tide may be propagated from the ocean through different channels [the Hainan Strait and the Malacca Strait] . . . in which case the same tide, divided

into two or more succeeding one another, may compound new motions of different kinds." Much later Thomas Young acknowledged
this hint in arriving at his principle of the interference of light.

Geomagnetism was a lifelong interest for Halley. Not only did he
theorize as to the nature and structure of the earth's magnetic field,
but the need to study the variation of the compass in different locations (partly with an eye to using it to measure longitude, a major
difficulty in Halley's day) led directly to his voyages over the entire
Atlantic Ocean. The isogonic lines on charts that resulted from the
voyages were long known as Halleyan lines. Later in life Halley came
to study the aurorae, and was likely the first to suggest they are somehow governed by what he called the "Magnetical Effluvia" of the earth.

It was Halley who established in 1686 the law of atmospheric pressure as a function of height above sea level, and who, after experi-
mentation with novel forms of transportable barometers, climbed Mt.
Snowdon in Wales ("a horrid spot of Hills, the like of which I never
yet saw") to put it to the test.

An interest in the energy balance of the atmosphere led to a theory
of tradewinds and monsoons, as well as the first comprehensive world
map of these. Such global considerations led Halley to suggest the first
scientific approach to determining the age of the earth: if the oceans
were originally fresh and their present salinity was due to the deposi-
tion of salts from rivers, then estimate the age of the earth from the
present salinity and its rate of increase. Halley fully understood the
limitations of the method and the difficulty of finding the rate of
change of salinity (wishing "that the ancient Greek and Latin Authors had delivered down to us the degree of the Saltness of the Sea,
as it was about 2000 Years ago"), but recommended the method to
posterity. For an age that had only two other alternatives—that the
earth was infinitely old or that the Scriptures were right with their six
thousand years—it wasn't a bad idea.

Halley, incidentally, was not one to be much trammeled by the
Scriptures. He lost the nomination for the Savilian Professorship of
Astronomy at Oxford in 1691 when Bishop Stillingfleet, "learning that
he was a skeptick, and a banterer of religion, scrupled to be concern'd;
'till his chaplain, Mr. Bentley, should talk with him about it. . . . But
Mr. Halley was so sincere in his infidelity, that he would not so much

as pretend to believe." The professorship went instead to David Gregory, which led to the story of a man appearing at a coffeehouse frequented by Halley and persistently asking to see him, explaining, "I would fain see the man that has less religion than Dr. Gregory!"

In truth, while Halley held liberal views on religion, for which Newton reprimanded him on occasion, he was not an atheist, and in 1704 succeeded to the Savilian Professorship of Geometry (despite grumbling from Flamsteed: "Dr. Wallis is dead: Mr. Halley expects his place, who now talks, swears, and drinks brandy like a sea-captain"). Oxford, in fact, saw fit to confer a doctorate on Halley in 1710. He was the recipient of more than one Oxford degree, even though as a student he had left the university before completing residence for a degree in order to undertake an expedition to the southern island of St. Helena. There he charted the far southern skies, thoughtfully naming a new constellation after the king: "Charles's Oak, deservedly translated to heaven in perpetual memory of King Charles II of Great Britain." Shortly after Halley's return from this expedition, "whereof [Charles] hath gotten a good testimony by the observations he hath made during his abode in the island of St. Helena," Oxford created him M.A. *per literas regias* (by virtue of the king's letters). Whatever else this may suggest, along with the fact that Halley had wangled free transport and lodging for the trip, it does suggest that had Halley lived in our age, he would have known how to write a grant application.

Halley's other work in the physical sciences (he had interests in the life sciences too) ranged from thermometry and calculations on the size of atoms to optics, he being the first to derive the formula for the focal length of a thick lens. His interest in optics led him to formulate a variety of intriguing questions that would give the average physics instructor today pause for thought. Questions such as, Why are media like glass and water transparent, while others are not? If light is slowed in passing through glass, what makes it resume a higher speed on emerging again?

Since his later professorship was in mathematics, Halley diligently pursued original research in that field too, publishing papers on logarithms, the finding of the roots of higher-order polynomials, and problems involving infinite quantities. As well, he produced scholarly editions of classical Greek geometry and translated from Arabic editions some of the works of Apollonius of Perga.

Halley was indeed a Renaissance man. He grasped the modern scientific method: "All that we can hope to do is to leave behind us Observations that may be confided in, and to propose Hypotheses which after Ages may examine, amend or confute." But his interests transcended science alone, and I shall return to him for a look at his nonscientific endeavors.

Candid Posterity and the Englishman. II

In writing of Edmond Halley in the previous chapter, I remarked on how poor posterity's memory has been in regard to that remarkable man. A century or so after his death in 1742, when some perspective had been gained, he was hailed as one of the greatest scientists of his time, perhaps second only to Newton, but now in the late twentieth century we associate him with the famous comet and little else.

It would be interesting to know what Halley himself might have regarded as the highpoints of his long life of eighty-five years. He led an extraordinarily full life, far beyond the range of most scientists. At one time he was a sea captain facing mutiny, he may have been a spy, he certainly advised on naval fortifications in the Adriatic, and there was the story of how Peter the Great of Russia gave him a ride in a wheelbarrow.

Curiously, though, we are not entirely sure just when he was born. He himself was sure, always giving his birthdate as 29 October 1656, but this implies that he was born only seven weeks after his parents' marriage in one of London's fashionable churches, which has seemed unlikely to later writers on the subject. Unfortunately, the Great Fire of London in 1666 probably consumed the official record of the event, much as it did many of his father's sources of wealth. We are as uncertain of Halley's financial circumstances as we are of his age; at times quite impecunious, he was yet able to finance the publication of Newton's *Principia* when only thirty years old.

On one thing all seem agreed, and that is Halley's personality. What a refreshing change the man is from all the tortured geniuses of history! His contemporaries write of him that "he always spoke as well as acted with an uncommon degree of sprightliness and vivacity . . .

First published in *American Scientist*, 74:55, January–February 1986.

he appeared animated in [the presence of his peers] with a generous warmth which the pleasure alone of seeing them seemed to inspire; he was open and punctual in his dealings; candid in his judgement; uniform and blameless in his manners"; and so on, almost ad nauseam. He even had a long and happy marriage: "Mr. Halley the following year 1682 changed his Condition, marrying Mrs. Mary Tooke, an agreeable young Gentlewoman; and a Person of real merit; she was his only wife, and with whom he lived very happyly, and in great agreement, upwards of 55 years; he had by her, that liv'd to grow up, one Son, and two Daughters."

This pleasing and well-adjusted personality was tried to the full by Isaac Newton. Yet without Halley's enduring patience and goodwill—indeed, without Halley's initiating the event that led to Newton's writing his masterpiece—the genius of Newton and all that flowed from it might never have been fully realized. As Richard Westfall in his recent biography of Newton puts it: "In August [1684], Edmond Halley traveled up from London to put a question that only Newton could answer. No other interruption stirred him so deeply. Halley's visit changed the course of his life."

Halley's question arose from a discussion he had had with Robert Hooke and Christopher Wren, and was essentially the question of how to derive Kepler's laws of planetary motion from dynamical principles. Abraham DeMoivre reports on the meeting: "After they had been some time together, the Dr asked him what he thought the Curve would be that would be described by the Planets supposing the force of attraction towards the Sun to be the reciprocal to the square of their distance from it. Sr Isaac replied immediately that it would be an Ellipsis, the Doctor struck with joy & amazement asked him how he knew it, why saith he I have calculated it, whereupon Dr Halley asked him for his calculation without any farther delay, Sr Isaac looked among his papers but could not find it, but he promised him to renew it."

So began the *Principia*. But it needed much of Halley's tact, persuasion, and at times flattery to bring it to fruition. Preliminary descriptions of Newton's work brought the quarrelsome Robert Hooke to the fore, and Halley had to advise Newton that "Mr Hook has some pretensions upon the invention of ye rule of the decrease of Gravity, being reciprocally as the squares of the distances from the Center.

He sais you had the notion from him." Newton, always paranoid over such matters, was thrown into a fit of white-hot rage, repeatedly threatening to leave the work unfinished rather than become embroiled in such charges. It took immense effort on Halley's part to smooth things over, although Hooke went to his grave still Newton's enemy. And then, to finish the episode, the Royal Society ordered "that Mr. Newton's book be printed, and that Mr. Halley undertake the business of looking after it, and printing it at his own charge." Nevertheless, Halley showed himself an accomplished poet by penning a set of graceful Latin hexameters in praise of Newton and prefixing them to the *Principia.*

Perhaps it was with relief that a few years later Halley found himself in company far removed from academia. If so, the relief was short-lived. He now became Captain Halley, a fully commissioned officer of the Royal Navy, and learned that "his Maty. has been pleased to lend his Pink the Paramour for your proceeding with her on an Expedition."

This expedition had its origins in the Royal Society and initially included a circumnavigation of the world, but it was later toned down to Atlantic travels only. Halley's prime interest was to see whether the deviation of the magnetic compass, known to vary with geographical position, could be calibrated as a measure of longitude. The lack of a method for finding longitude at sea was the bane of contemporary navigation. Halley's being made the actual captain, rather than merely scientist-on-board, was to ensure that no one would overrule his wishes en route.

Their Lordships at the Admiralty meant to make the expedition pay; included were orders "to stand soe farr into the South, till you discover the Coast of the Terra Incognita," to observe the coasts of West Africa and South America, and "to visit the English West India Plantations . . . to lay them downe truely in their Geographicall Scituation."

Regrettably, His Majesty's pink, all sixty-four feet and eighty-nine tons of her, lacked seaworthiness: "During the bad weather on Sunday, her streining opened some leaks which are considerable for a new shipp, and have discovered an evill wee did not foresee; for having only hand pumps, and our ballast being Sand, the bilge water with the motion of the shipp brings the Sand to the pumps and choaks them." After a two-week delay for caulking and a change of ballast, the expedition sailed in late November 1698.

Another important delay befell them at Madeira. "By reason of the Holydays it was not possible for the Shipps to have their Wines on board before this day, wch occasioned the Admirall [briefly accompanying Halley] to leave the Island the same night he arrived, being unwilling to waite so long."

The *Paramour* (or *Paramour Pink,* as Halley delightfully called her—pink really being the type of ship) reached the "Cape de Virde" Islands in early January, there to find two English ships, "one of which . . . was pleased to fire at us severall both great and small shott. We were surprized at it." The resulting altercation over just who was thought to be a pirate being straightened out, Halley found reason to give up the Antarctic leg of the voyage: they would likely not reach there before the southern winter. They did, however, proceed to Brazil ("the Portuguez were very willing to find pretences to seize us") and up to the West Indies.

En route Halley began to have serious trouble with the crew. He discovered the boatswain deliberately steering so as to miss an island Halley wished to visit, then the officers "showed themselves uneasy and refractory," and finally, off Newfoundland, there was "the unreasonable carriage of my Mate and Lieutenent, who made it his business to represent me, to the whole Shipps company, as wholy unqualified for the command their Lopps [Lordships] have given me. He was pleased so grosly to affront me, as to tell me before my Officers and Seamen on Deck . . . that I was not only uncapable to take charge of the Pink, but even of a Longboat." The crew sneered at Halley's amateurism, and only his swift action in clapping the mate below decks prevented mutiny. With satisfaction Halley sailed the *Paramour* home himself, although in the subsequent court-martial he was incensed that the crew was merely reprimanded and the matter dismissed as "only some grumblings such as usually happen on board small Shipps."

Much to his credit, though, within a month or two of this the *Paramour* had been refitted, a new crew appointed, and Halley cheerfully committed to reaching the Antarctic seas. Typically, he succeeded. There were, of course, problems along the way ("I was Obleged to putt into Ryo Jennero in Brasile to gett some Rumm for my Ships company"), but on 1 February 1700 they were below latitude fifty-two degrees south and "fell in with great Islands of Ice, of soe Incredible a hight and Magnitude, that I scarce dare write my thoughts of it." Soon "we were in Imminent Danger to looss our ship and lives, being Invironed wth Ice on all Sides in a fogg soe thick, that we could not see it till was ready to strike against it."

I won't dwell on the adventures homeward bound, although they included a brief imprisonment at "Fernambouc in Brassile" under suspicion of piracy. The idea of finding longitude from the compass, of course, failed, although the expeditions did provide the first wide-scale mapping of isogonic ("Halleyan") lines. Other scientific endeavors, including observations of flora and fauna, were more successful.

Halley's seafaring days were not yet over. In 1701 the *Paramour*, under his command, was sent to make a detailed survey of the English Channel. It was somewhere about here in the story that Halley seems to have been involved in spying on French ports. I have not been able to track down any details, but a friend of his writing soon after Halley's death hinted strongly at this. If there was an official cover-up, it has lasted well. Halley was in any case something of an expert in gunnery, and at one time and another did original research in the theory of ballistics. His knowledge of fortifications was extensive, and led in 1702 to his crisscrossing Europe (dining with emperors and princes en route) on a diplomatic mission to rebuild the fortifications of Trieste on the Adriatic Sea.

Halley's naval experiences predated his command of the *Paramour* by at least ten years, for he had become interested in undersea diving as early as 1688. As Halley first found it, the diving bell suffered for want of a method to change the air in it while it remained underwater, a problem aggravated by the candles or lanterns used in it for lighting. Halley devised an ingenious way of supplying air from barrels through leather hoses, and soon was himself diving to depths of over sixty feet for periods of more than an hour. He went on to invent means whereby, wearing a "girdle of leaden shott" and a "cap of Maintenance," he could leave the diving bell and walk across the seabed. So successful was all this that he formed a commercial salvage company to work on shipwrecks, the shares of which were quoted in London through much of the 1690s.

Halley's early reputation for naval knowledge led to his brief, if seemingly tumultuous, friendship with the Russian czar known to history as Peter the Great. This larger-than-life individual had arrived in England in 1697 to learn at first hand the arts of shipbuilding. He would work as a laborer in the shipyards of Deptford, but it was Halley who was sent for to be his guide and mentor.

Peter, then in his twenties, was gargantuan. A bear of a man, said to have been seven feet tall, he was of herculean physical prowess. All his qualities (says the *Encyclopaedia Britannica*) were on a colossal scale. "His rage was cyclonic: his hatred rarely stopped short of extermination. His banquets were orgies, his pastimes convulsions. He lived and loved like one of the giants of old." One wonders how Halley viewed his assignment. Still, a contemporary reports that Peter found Halley "equal to the great character he had heard of him . . . and ranked him among the number of his friends." And so arose the story, unauthenticated, that one such convulsive pastime included a wheelbarrow race among friends in which the giant Peter drove his passenger Halley through a holly hedge. What is authenticated is that Peter and his entourage during their stay very nearly destroyed the fine mansion, Sayes Court, lent to them by its owner. The first item in a long list of damages was three hundred broken windows.

Well, I must draw the story of Halley to a close, although there is much more to be said of him. His interesting times as comptroller of the Chester Mint, his laying the foundations for the theory of annuities and life insurance—complete with cheerful Halleyan philosophical admonishments: "How unjustly we repine at the Shortness of our Lives, and think ourselves wronged if we attain not old Age; whereas it appears that the one half of those born are dead in Seventeen Years Time."

Halley himself certainly did not repine at any shortness of his own life. We find him, well into his eighties, still astronomer royal at Greenwich, just concluding his nineteen-year research program of lunar observations. He still went up to London by river each week to dine with his friends, although by then "Dr. Halley never eat any Thing but Fish, for he had no Teeth." He continued astronomical observations to within a few months of his death, when a "paralytic disorder gradually increasing, and thereby his strength wearing, though gently, yet continually away, he came at length to be supported by such cordials as were ordered by his Physician, 'till being tired with these he asked for a glass of wine, and having drank it presently expired as he sat in his chair without a groan on the 14th of January 1742 in the 86th year of his age."

He was buried in St. Margaret's churchyard, not far from Greenwich, where his wife had been buried six years earlier. His tomb includes the inscription "Astronomorum sui seculi facile princeps" (unquestionably the greatest astronomer of his age). Sadly, a biographer, Angus Armitage, writing in 1966, reports that "today the old village church is a ruin standing beside a busy highway and the tomb is overgrown with brambles and thistles." This Englishman surely deserved better at the hand of candid posterity.

Bloody Sirius

As no doubt the sales of Agatha Christie's books testify, everyone loves a good mystery. In astronomy we always keep a few good mysteries knocking about in case a cloudy night leaves time on our hands. One may ponder just how much astronomy the Stonehenge people really did build into their monument, or, at the other end of history, the nature of cosmic wormholes as tunnels to other universes. Some mysteries, for example the question of whether Mars has canals, are eventually resolved, but the old favorite among astronomy mysteries has been around for well over two centuries and happily looks to be with us always. I refer to the vexing case of the star Sirius.

In 1760 Thomas Barker, writing in the *Transactions of the Royal Society*, noted that while there was generally good agreement between the classical Greeks and modern observers as to the appearance of prominent stars, a glaring discrepancy was to be found in Sirius. While even the most casual observers today would agree that Sirius is white, Barker pointed out that well-known figures of the past, notably Horace, Cicero, Seneca, and particularly Ptolemy, had referred to it as red. The question of whether or not Sirius has changed color so drastically on so short a time scale, astronomically speaking, has exercised astronomers ever since, and they are still furiously digging up evidence on both sides of the argument.

What has so intrigued astronomers about this is that astrophysics itself tells us that Sirius must once have appeared red. The bright white star we now see has a tiny companion that is near the end point of its evolution. Once the more massive of the two stars, it evolved much faster, and in the course of that evolution became a red-colored giant star. The tiny dwarf, in fact, is but the core of that red giant,

First published in *American Scientist*, 77:429, September–October 1989.

revealed when the outer layers of the giant were lost to interstellar space. As a red giant, however, its light would have swamped that of its white partner, so to the unaided eye the unresolved pair would have appeared red. Moreover, not only would Sirius have appeared red then; it would have appeared much brighter than it does now.

The controversy thus continues on two fronts. The first is whether the historical record is right, both as to the putative redness of Sirius *and* its apparent brightness. The second is whether a plausible astrophysical explanation can be found for such evolution in so short a time, if indeed there's really anything to explain.

The historical record is the first stumbling block. It's difficult to appreciate just how slim is our understanding of who said what among the ancient Greeks, let alone why. Norriss Hetherington, in his recent book *Ancient Astronomy and Civilization,* gives some flavor of the problem: "Eudoxus . . . attended Plato's lectures. Upon his report of what Plato said a string of statements followed, each based upon a previous statement. Eudoxus' report is lost. However, it was summarized by Eudemus in his own *History of Astronomy.* This work also is lost. It was commented upon, though, by Sosigenes. . . . Sosigenes' work is lost too. It was used, though, by Simplicius of Athens in the sixth century A.D. . . . Simplicius' work survived." Thus, through lost reports spanning nearly a thousand years, do we glean something of those lectures in the Academy. What reliance can we place on Simplicius' report?

Even Ptolemy, of whose life we know nothing but who left us the best-known of ancient astronomical texts, the *Almagest,* comes to us via a tortuous route. Centuries after Ptolemy's death, the *Almagest* passed to the Muslim world, where it enjoyed wide circulation in Arabic translation and was hand-copied time and again over many centuries before finding its way to medieval Europe and so eventually to us. How far should we trust even the best translations of today?

Certainly the most recent translation, that of Toomer, presents some interesting aspects. For one thing, Sirius isn't the brightest or even the second-brightest of stars in the catalog of the *Almagest;* it's the third-brightest after Betelgeuse and Regulus, although elsewhere in the *Almagest* it is referred to as "the brightest of [the fixed stars]."

In figure 4.1, I have plotted the *Almagest* magnitudes of about a

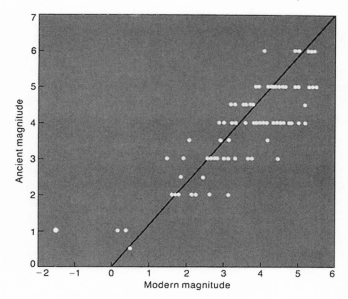

Figure 4.1 *The magnitudes of about a hundred stars as reported in Ptolemy's* Almagest *are plotted here against currently accepted values. A perfect match would fall on the diagonal line. The larger dot on the lower left, representing Sirius, shows a disagreement of nearly two magnitudes. (Original diagram by Susan Hochgraf,* American Scientist.)

hundred stars in the celestial neighborhood of Sirius against their modern values. Sirius is the larger dot on the lower left, where it certainly stands out from the trend. Why did Ptolemy call it first-magnitude when today it is some two or three magnitudes brighter? When I once remarked on that, a critic replied that the reason was simply that Ptolemy had no concept of zero or negative numbers. That, I would think, is a moot point, but even if true it begs the question: Why not shift the whole scale down a few digits and put Sirius in a class by itself, Betelgeuse and Regulus notwithstanding?

In any case, on the question of whether or not Sirius was *brighter* two thousand years ago (a question not discussed much in the literature), the historical record, taken at face value, has a negative answer. If anything, it might have been fainter, but don't bet much more than the proverbial wooden nickel on it.

As to the color of Sirius, the historical record seems more definite. The *Almagest* mentions the colors of only six stars out of the 1,022 in its catalog. Those six are Arcturus, Aldebaran, Pollux, Antares, Betelgeuse, and Sirius, all six being called "reddish." Now as your friendly neighborhood astronomer, armed with his *Bright Star Catalog*, can tell you, the first five are *really* red; there's no question that when Ptolemy said "reddish," he meant red, not yellowish or orangish or anything like that, but bloody well *red*. Add to this the laborious diggings of one T.J.J. See, who finds that Seneca notes "the redness of the Dog Star is deeper, that of Mars milder," that Pliny in his *Natural History* groups Sirius, Mars, and the rising sun as being "igneous," and that Homer, Hesiod, Horace, Virgil, and others all allude to the redness of Sirius, and you begin to think you're onto something. Now Thomas Jefferson Jackson See was not one exactly vested with ultimate authority on matters astronomical, but should he be dismissed as an ignoramus on classical translations too? We need the view of a modern scholar of classical Greek.

The whole question of Sirius's past color hotted up recently with the publication of a paper in *Nature* by Schlosser and Bergmann reporting an eighth-century European reference to a red Sirius that would have been independent of any Greek tradition and less liable to unknown factors involving hearsay. These authors also pointed to Babylonian cuneiform texts calling Sirius red. The Schlosser-Bergmann paper followed a note by Pennick drawing attention to early Icelandic references to Sirius as being red. These, of course, place even more stringent limits on the time scale for the companion star to evolve from a red giant to its present status of a tiny white dwarf.

Some have tried to escape this increasingly uncomfortable situation by appealing to the fact that Sirius played a central religious role for the early Egyptians (some of whose works carried over into classical times), inasmuch as the time of year when it could be seen to rise just ahead of the sun coincided with the flooding of the Nile. And a bright star, like the sun itself, when seen close to the horizon will appear much redder than normal. Thus references to a red Sirius are taken to refer not to the star hanging in the winter evening sky but to its all-important rising in a summer's dawn, herald of a new season of growth. Well, I don't know. Maybe the Egyptians would think that a big moment, but would the much later Greeks and Romans subscribe to it? And almost certainly not the very much later central Europeans and Icelanders. Wouldn't somebody have made passing reference to the winter Sirius? And what about other cultures, those of the Orient or Middle East? Did their early Sirius-watchers have anything to report? I'm sorry to say the answer is yes!

That's the trouble with mysteries. Just as you're wallowing luxuriously in their murky depths, somebody pulls the plug. Thus Ian Ridpath, without even a word of apology, reports that the Roman poet Manilius, writing around 15 C.E., says of Sirius that "the beams it launches from its sea-blue face are cold." Bicknell, however, replies that Ridpath has got it wrong and that Manilius was in fact speaking of a *red* Sirius. But before Ridpath, Tang reported that Chinese records of the first century B.C.E. contain a passage beginning, "The white is like Sirius, the red like Antares." On the other hand, the same records announce Betelgeuse to be yellow instead of its present deep red, rather muddying the waters. But then comes Robert van Gent, reporting with positive glee nearly a dozen additional early references to a white Sirius, ranging from Spaniards to Chinese and from ninth-century-B.C.E. Persians to fourteenth-century-C.E. Geoffrey Chaucer. I hope these spoilsports are satisfied. Like a juror listening to psychiatrists called by the defense at a murder trial to testify against their counterparts called by the prosecution, what is one to make of it all? I'm inclined to think the case not yet entirely closed. At least I hope not!

What of modern astrophysics? Can it handle the possibility of a recently red Sirius? Well, we romantics are taking it on the chin a bit there too. The Schlosser and Bergmann paper stimulated considerable interest among stellar evolutionists, some of whom have since bent over backward to accommodate a historically red Sirius. But only with limited success.

If it were just a question of changing Sirius's color quickly, we could perhaps squeeze by. Schönberner has produced theoretical models of red giant stars that strip themselves to their white-hot cores in something like a thousand years, the time scale being very dependent on mass. The trouble comes in getting the hot core to cool to the current paltry forty-eight thousand degrees Fahrenheit of the Sirius companion in anything much under ten or a hundred million years.

Much the same is true of getting the luminosity down to the observed level. Then more trouble comes with the question of how to get rid of the giant's outer layers so quickly. It's the classic problem of the mystery genre—what to do with the body?

There is certainly no obvious nebula or other debris surrounding Sirius that might be so interpreted, and Bruhweiler, Kondo, and Sion have made a careful search of the Sirius region with the *International Ultraviolet Explorer* satellite without turning up any debris much more than a hundred-thousandth of a solar mass. But, like the experienced mystery writer, these authors have an ingenious solution to all the difficulties. The dwarf companion, they propose, did in fact emerge from a red giant only some tens of millions of years ago, allowing plenty of time to cool and wane and have the debris depart the scene. Then, much more recently, the dwarf suffered a sudden puffing up into a red-giant-like configuration caused by a thermonuclear runaway as a very thin outer layer of hydrogen diffused into the helium layers below to encounter upward-extending carbon. Because of a lack of fuel, however, the configuration rapidly collapsed after a few hundred years, returning the dwarf to the continued quiet cooling and fading we see today.

Incidentally, Bruhweiler, Kondo, and Sion also bring up an old favorite among Sirius apologists: the dust theory. At least as early as Sir John Herschel in the mid-nineteenth century, people were talking about an interstellar dust cloud drifting across the line of sight between us and Sirius, the dust reddening the light of the star. But as Bruhweiler et al. point out, even in modern form this theory produces a difficulty opposed to that of the red giant theory: in the latter Sirius would have appeared much brighter than now, while in the dust theory Sirius, its light dimmed as well as reddened by dust, would have appeared much fainter. And how to get rid of the dust so quickly?

Joss, Rappaport, and Lewis have an interesting go at the problem by proposing that the binary nature of the Sirius system is important. They point out that the currently rather well separated pair of stars would once have seemed otherwise when the now dwarf companion was a giant. Then gravitational interaction could have had serious consequences for the giant's outer layers, leading to what is called critical lobe overflow, which would make the giant's evolution quite different from the case of an isolated star. Final stripping of any residual envelope would happen within a few orbital revolutions (eighty years apiece), leading to very rapid changes. As to the debris problem, at least some of the lost envelope would pass to the other star rather than circumstellar space. These authors, though, are honest enough to admit their hypothesis to be "an intriguing but probably unsatisfying explanation for the reported transition of Sirius."

So there we are. Conflict among the historians and nothing very satisfactory from the theorists. Just the kind of thing science thrives on. I'm sure we haven't heard the end of it yet.

Stonehenge and the Archaeoastronomers

Archaeoastronomy, the application of astronomy to archaeology, has been around for more than a century. Today it spans a wide spectrum of problems, from the astonishing calendars and ruined jungle cities of the Maya, through rock structures and paintings of the American Southwest and the medicine wheels of Wyoming and Alberta, to the navigational methods that brought people to the islands of the South Pacific and putative Scottish observatories that five thousand years ago utilized distant mountain peaks as optical sights for tracking the sun and moon. Nothing, however, epitomizes archaeoastronomy as does its application to Stonehenge.

Close to the busy highways of southern England, Stonehenge is well known and accessible, yet it remains one of the most mysterious places on earth. It is a casual observer indeed who feels no thin shiver of thrill on the spine when approaching by car or bus and seeing that strange wreckage of rocks rising from Salisbury Plain. And so it has long been. There are references to Stonehenge as far back as the twelfth century; even then, people wondered about its origin and purpose. Legends and myths about Stonehenge, their darkness often exceeded only by their wildness, have abounded ever since. (A less-certain reference to Stonehenge is attributed to Hecataeus of Thrace in the fourth century B.C.E.)

Stonehenge is largely composed of a series of concentric rings, some marked by stones, some not. It does, however, have an axis, best delineated by a straight avenue that evidently served as an entranceway, and here astronomy first found its foothold on Stonehenge. John Aubrey, who opened the modern era of Stonehenge studies in 1666, may have known that the axis is aligned toward midsummer sunrise.

First published in *American Scientist*, 78:103, March–April 1990.

Certainly, William Stukeley drew attention to it in 1740 (along with an enduring but totally false theory that Stonehenge had been built by Druids).

In 1900, the noted astronomer Sir Norman Lockyer used this alignment, or near alignment, as the basis for an attempt to date Stonehenge. His idea was that although the current alignment of the avenue was not exact, it had been exact when the avenue was laid down. Presumably, the difference accrued over time from the fact that the direction of the rising midsummer sun depends on the 23½-degree angle between the plane of the earth's equator and the plane of the earth's orbit about the sun. This angle, designated ε, is known to oscillate slightly over a long time scale, thus changing the direction of sunrise. Lockyer reasoned that he could find the change in ε from the present alignment error, and since astronomers knew the rate of change of ε, he could calculate the time elapsed since the avenue was built.

Lockyer should have known better. To begin with, the change in the direction of sunrise is exceedingly slow—less than a degree over several thousand years. In addition, the avenue was probably not laid down with a precision that would be high by modern standards; in any case, the weathering of millennia has degraded its outlines considerably. Even if appropriate standing stones were used to determine the axis, their bulk and possible slumping would produce an ill-defined line, and an uncertainty in the direction of the original axis of at least 0.1 or 0.2 degrees. Worst of all, did sunrise refer to the moment when the upper or the lower edge of the sun touched the horizon? Results obtained using these two definitions of sunrise differ by four thousand years. And incidentally, was the horizon then what it is now?

So, although Lockyer's date of 1680 B.C.E. for the building of Stonehenge—1840 B.C.E. if we use modern numbers—wasn't all that bad, as it turned out, his work nevertheless met with skepticism even then and with considerable scorn from archaeologists today. Lockyer, it must be admitted, seems to have had a certain arrogance. He was the founder and first editor of the journal *Nature*, which led an annoyed contemporary to remark that the editor of *Nature* too easily confused himself with the Author of nature.

Lockyer's name is also associated with the interpretation of another feature of Stonehenge: a rectangle set upon an outer circle of holes and originally defined by four large stones. By as early as 1846 it had been discovered that the shorter sides of the rectangle are parallel to the monument's axis, and so point toward midsummer sunrise in one direction and midwinter sunset in the opposite direction. Lockyer noted that the diagonals of the rectangle point to sunrise/sunset directions on dates that would, along with dates of the solstices, divide the year into eighths. He therefore suggested that the rectangle served as a calendar. However, alignments on the equinox sunrises would be needed to complete this picture, and these were missing, so again Lockyer's claims were dismissed.

Astronomers seem to have retreated in face of the onslaught against Lockyer, and Stonehenge was left to the archaeologists for more than a half-century. Then in the mid-1960s the astronomers returned with theories about Stonehenge that caused near apoplexy in some archaeologists. The echoes of the ensuing clashes have not entirely died away to this day.

These theories began with the work of two men: C. A. Newham, a retired manager of a gas board in England and an enthusiastic amateur archaeoastronomer, and G. S. Hawkins, a British-born astronomer from Boston University and the Harvard-Smithsonian Observatory. It is a small but fascinating case history to see how Newham, the amateur, fared in the professional science arena compared to Hawkins. Newham, whose conclusions were a subset of Hawkins's conclusions, was limited in part by having only logarithmic tables; Hawkins had access to a mainframe computer.

Newham would later recount how he had found it impossible to get his work published in scientific journals, including *Nature* (some irony there!). He appealed to the leading archaeological authority on Stonehenge, who apparently gave him to believe his work was not without merit. But there was still no publication beyond an article in the *Yorkshire Post* in March 1963, which drew no iota of professional interest. One can imagine Newham's feelings when later that year a paper by Hawkins appeared in *Nature* and produced a rush of radio and television interviews, including an interview with the leading archaeologist, who somehow failed to mention Newham. (It must be said that these wrongs were later righted and Newham duly credited, although that might not be the word of archaeological choice.)

What Newham and Hawkins had done was to introduce a new,

more complicated astronomical factor into Stonehenge studies, namely the motion of the moon. To explain briefly: The earth orbits the sun in a year, during which time sunrise appears near the northeast at midsummer, moves to the southeast by midwinter, and returns again to the northeast. Thus in one year there are two points of extremity for sunrise. The moon orbits the earth in one month, and during that time its rising point also covers an arc of the eastern horizon, moving from roughly northeast to southeast in two weeks and then back again. However, the moon's orbit precesses, or wobbles, in an 18.6-year cycle, during which the length of the arc changes considerably. At one point of the cycle, the arc may extend from only a little north of east to a little south of east, whereas nine years later the arc may run from north of northeast to south of southeast. There are thus four extreme rising points for the moon over an 18.6-year cycle.

Referring to an accurately scaled plan, Hawkins entered the positions of a good many of Stonehenge's stones and holes in a computer. He also entered equations for computing rising and setting positions of the sun, moon, and a number of bright stars and planets. Long-term effects were allowed for, as were various minor ones. The computer was programmed to determine various rising and setting directions for these bodies in about 1500 B.C.E. and to compare the results with alignments defined by pairs of prominent stones and holes at Stonehenge.

Hawkins's program found no alignments with stars and planets, but produced no fewer than ten alignments for the sun and fourteen for the moon. In particular, the diagonal of the rectangle mentioned earlier was found to point toward one of the moon's extreme positions and the long side of the rectangle to another of the moon's extreme positions. Hawkins estimated the chance of these various alignments being accidental as less than one in a million. He concluded that Stonehenge had been built, anthropological reasons aside, as some kind of observatory.

The rectangle, incidentally, has also been involved in another astronomical claim, namely that only at the latitude of Stonehenge will the claimed alignments produce a rectangle. If the construction were moved farther north or south, the figure would "degenerate" into a parallelogram. The implication is that the neolithic builders knew enough to deliberately select the latitude for this reason.

While no doubt Hawkins did not see it as such, all this was a bit of a preemptive strike against the archaeologists. The whole idea and approach, dependent on mathematical astronomy, were foreign to their own methods, and they were ill-equipped for launching any claims of error in this work of Hawkins. But hate it they did, at least some of them, beginning with the very title of Hawkins's paper: "Stonehenge Decoded." When one has spent one's career working painstakingly toward the solution of a problem, one does not take kindly to an outsider appearing with the airy announcement of a solution—so that everyone can now go home. When Hawkins's paper was later expanded into a popular book of the same title, the fury of the archaeologists only intensified.

This was also true when Hawkins's second paper appeared in the 27 June 1964 issue of *Nature* under the title of "Stonehenge: A Neolithic Computer." In this paper Hawkins focused on a different aspect of Stonehenge: the outer ring of holes, known as the Aubrey holes. Relatively shallow and wide, these apparently never have held large stones; instead, remains of cremations suggest some ritualistic purpose. Hawkins attached significance to the fact that there are fifty-six Aubrey holes and that $3 \times 18.6 = 55.8$. The 18.6-year lunar cycle, in turn, is related to eclipse occurrences, and eclipses are known to have been important omens in early cultures. Were the Aubrey holes part of some arrangement for predicting eclipses? Hawkins worked out an elaborate scheme whereby the systematic shifting of markers around the Aubrey circle "will predict accurately every important lunar event for hundreds of years." Indeed, "a complete analysis shows that the stone computer is accurate for about three centuries," after which a minor adjustment would reset it. The question, though, was whether a people who had not yet invented writing could have evolved so elaborate a computer; also, was there any compelling reason to believe they had? Archaeologists thought not.

Fred Hoyle, a leading British astronomer and no laggard when it came to novel ideas, suggested an alternative scheme by which the Aubrey holes could be used to predict eclipses. This too involved moving markers around the circle, which now was seen as a representation of the paths of the sun and moon in the sky. The sun marker would be moved so as to make one circuit each year, the moon marker, one circuit each month, and markers for the nodes of the moon's orbit, one circuit each 18.6 years. Eclipses could be predicted when it was seen that appropriate markers would soon be opposite or coincident with one another. The choice of fifty-six for the number of holes, however, was not very felicitous, and in any case Hoyle's scheme was all very ad hoc and unconvincing.

Hoyle had other astronomical ideas for Stonehenge. There was the question of why the Heel stone, lying outside the main body of the monument, is close to but definitely displaced from the point of midsummer sunrise as seen from the monument's center. Hoyle suggested that if one wished to determine the date of the solstice, it would be best not to put a marker precisely at the midsummer sunrise point, because the change from day to day just then is virtually indiscernible. It would be better to place it somewhat to the south, and note both the day prior to the solstice, when sunrise is at the marker, and the day after the solstice, when sunrise passes the marker. The solstice itself would lie midway between these two dates; in future years one would know that the solstice occurs a certain number of days after the first passing of the marker.

Archaeologists remained unconvinced, and in some cases were severely hostile to these schemes. There were statistical arguments over how much uncertainty one could tolerate in alignments and so forth, but the most telling objection was that the astronomers had simply ignored so much archaeological evidence. For instance, other neolithic monuments have holes very similar in shape and content to the Aubrey holes, yet none happen to number fifty-six. What price Hawkins's theory then? Stonehenge, like Rome, was not built in a day. Current estimates, mostly from radiocarbon dating, place its earliest stages at 3100 B.C.E., and building and rebuilding continued sporadically until at least 1000 B.C.E., during which time it came under the jurisdiction of various cultures. It is difficult to imagine astronomical plans being carried through such spans of time and circumstance.

Not much has happened by way of astronomical theories about Stonehenge since the uproar of the 1960s. Astronomers and archaeologists have learned valuable lessons and are now far more cooperative and appreciative of each other's views—and archaeoastronomy has moved on to other areas. As for Stonehenge itself, the British archaeologist Aubrey Burl has summed it up well: "Wrecked in antiquity, chipped for mementoes in the eighteenth and nineteenth centuries, threatened with demolition in the 1914–18 war, sold at auction in 1915 for £6,600 . . . this ravaged colossus rests like a cage of sand-scoured ribs on the shores of eternity, its flesh forever lost. Stonehenge grudges its secrets. Each one explained—the date, the source of the stones, the builders—leads to greater amazements, a spiraling complexity that even now eludes our understanding so that our studies remain two-dimensional and incomplete."

Alexander Thom and Archaeoastronomy

Although it is Stonehenge that springs to mind whenever there is talk of the astronomy of prehistoric peoples, archaeoastronomy for decades was dominated by someone who had little to do with Stonehenge. His name was Alexander Thom. He died in 1985 at the age of ninety-one, and in recent years the archaeoastronomical community has been assessing his successes and his failures.

Thom was an unlikely figure in this setting. He was an engineer and hydrodynamicist; in the latter capacity he was a senior official in Britain's Royal Aircraft Establishment during World War II. After the war he held the Chair of Engineering Science at the University of Oxford. And although he had astronomy as a hobby, it was largely directed toward grinding and polishing mirrors for small telescopes. ("My mother," writes Thom's son, Archie, "was greatly dismayed with the jeweller's rouge transferred from the seat of his pants to the cushions of the house furniture.")

Thom hailed from northern Scotland, a land he dearly loved, and, as an expert yachtsman, he spent summers cruising the coastal waters among the remote islands of the Orkneys and Hebrides. In these high latitudes the seasonal changes in the diurnal motions of the sun and the moon are very pronounced, and it is a natural setting for the kind of observations that play an important part in archaeoastronomy. A few years before his death, Thom recalled a 1933 cruise with his son and friends, during which they anchored at East Loch Roag in the northwestern part of Lewis Island in the Outer Hebrides. "As we stowed sail after dropping anchor," he wrote, "we looked up, and there, behind the stones of Callanish, was the rising moon. That

First published in *American Scientist*, 78:406, September–October 1990.

evening . . . we went ashore to explore. I saw by looking at the Pole Star that there was a north/south line in the complex. This fascinated me, for I knew that when the site was built no star of any magnitude had been at, or near, the pole of the heavens. . . . The Outer Hebrides have a charm of their own, not the least being their remoteness. To realize that megalithic man had once lived and worked there . . . aroused my interest in the workings of his mind." Callanish is one of the preeminent megalithic sites in Britain, and here, it seems, Thom's interest in matters megalithic was first fired.

Initially, though, his interests centered more on the metrology, rather than the astronomy, of prehistoric Britons. Britain, and indeed Ireland and the western littoral of Europe, have a surprisingly large number of prehistoric stone rings. One tends to think only of the famous few, such as Stonehenge, but there are thousands of rings, ranging from modest, simple circles a few meters in diameter to the huge complexity of Avebury with its giant stones laid out in circles with diameters of a thousand feet or more.

Prior to Thom, it seems, attention had been focused mainly on individual sites of particular interest. It is difficult, though, to deduce the intentions of the preliterate mind by studying one site that has weathered four or five thousand years. Thom realized that much might be gained by a statistical study of many rings. What could be learned from features the rings might have in common?

So Thom, at times aided by Archie, himself an engineer, started out by systematically surveying several hundred sites with previously unachieved accuracy. All other workers in the field acknowledge the precision of the Thoms' surveys. But it was the conclusions they drew from these surveys that really awoke their colleagues.

An early question posed in one of the studies was whether the rings had been laid out using a common unit of length. Thom concluded that they had, a length that he set at 2.7 feet and dubbed the megalithic yard. Moreover, from a similar study of what are called cup-and-ring markings carved into the faces of some of the megaliths, he deduced the existence of a megalithic inch that measured one-fortieth of the yard, and later a megalithic rod of 2.5 megalithic yards. (The significance of the ringed cups remains entirely unknown.) The most startling result, however, was the possibility that these units prevailed wherever rings were found. That a prehistoric society, at times thought of as comprising nomadic hunting bands, should have shown such uniformity across such spans of time and place was something to contemplate. On the other hand, those were the same people who had joined in constructing the great monuments of Avebury, Silbury Hill, and Stonehenge, efforts that in terms of the population's resources far exceeded the expenditure of the United States in putting a man on the moon.

Hardly less intriguing was Thom's study of the geometry of the stone rings. To the casual observer the rings are approximately circular, but some come closer to a perfect circle than others. One would presume either that the builders didn't care much about precision or that the rope compasses they used to lay out a circle were liable to stretch. What emerged from Thom's work, however, was that where noncircularity is present, it was deliberate. Thom found a handful of shapes among the rings: true circles, ellipses, and what were called flattened circles and eggs. The latter in particular are interesting, being based, according to Thom, on right triangles, with the implication that the designers knew Pythagoras's theorem, even though they lived millennia before the Pythagoreans. Thom also showed how the flattened circles might have been systematically constructed using particular radii in the geometry. It was surmised that the reason for these curious shapes was early man's discovery that a true circle cannot have an integral number of units in both its radius and its circumference, whereas the other shapes are much better approximations to that condition. As to the ellipses, an excellent example was the innermost "horseshoe" ring at Stonehenge, whose major axis the Thoms found to be aligned within arc minutes of the monument's main axis.

This interest in megalithic geometry led to a study of Carnac, a site on the Atlantic coast of France. Here is the largest of all known menhirs (monumental standing stones), Er Grah or Le Grand Menhir Brisé, which, though now fallen and broken, stood over sixty-five feet tall and weighed well over three hundred tons. The shaping, transportation, and erection of such an object by the engineers of the day staggers the imagination. (One can too easily dismiss the task as impossible. It has been shown that such feats, although immensely difficult, were within reach of the simple tools of the day, given enough manpower. I had the latter point emphasized on an exam once, when I asked my students to explain the word "menhir"; one wiseacre replied that when they were struggling to put up Stonehenge, one worker turned to another and gasped, "We need more menhir.")

Er Grah may have served as a foresight for observers standing kilometers behind it, watching sunsets or moonsets on the sea horizon. (As on a rifle barrel, an observer aligns a remote object along two sights; the distant sight is called the foresight, and the nearer sight is called the backsight.) The question of backsights focused attention on one of the most puzzling of megalithic features, great rectangular or fan-shaped arrays of smaller stones laid out on grids of a few feet spacing covering thousands of feet of terrain. They are not unique to Carnac, being found also in England and Scotland, but those near Carnac are the largest. With an egglike ring at each end, they are enigmatic indeed. It was suggested that they served as a precise interpolating device, the observer moving among the stones until finding one from which Er Grah pointed precisely at the moonset on a given night. A succession of such points night by night would allow the deduction of the moon's northernmost position, even though it did not attain that position at the moment of moonset. Thus the appellation "stone graph paper."

The idea of alignments where backsight and foresight might be separated by miles became of increasing interest to Alexander Thom. He and his coworkers suggested that at Stonehenge, for example, there were alignments based on foresights eight or more miles away pointing in astronomically significant directions. Again, many surveys were carried out, particularly in the higher latitudes of Scotland, where even slight changes in solar or lunar sky coordinates bring dramatic changes in rising and setting directions. Often the foresights were natural features of the horizon—clefts in a skyline of distant mountains, for instance—but the backsights were (with a few singular exceptions) less convincing.

Still, statistical analyses were carried out and impressive results claimed. The moon's motion is complex and can require decades of observation to unravel, yet Thom presented evidence that the ancient observers had known its subtleties to resolutions of an arc minute or so. They might even have been involved with such niceties as determining the setting direction of the edges of the moon when the moon's distance and hence its apparent size vary slightly.

How has all this stood up? As an outsider scanning the literature on the subject, I'd have to say not too well. A collection of papers written in memory of Thom was published a couple of years ago. All the authors praise his hard work, the reliability of his surveys in the field, and his pioneering spirit, and no one impugns his honesty or sincerity, but there appears to be skepticism about many of his conclusions. The difficulty seems to lie chiefly in that all-too-often vexatious area, the statistical analyses. Thom was given to using phrases like "adopting only the most reliable cases" without making clear what criteria (other than personal judgment) were used to distinguish reliability, which of course could lead to the introduction of unconscious biases. And even though the notable statistician D. C. Kendall independently gave the nod to the megalithic yard, recent analyses of newer data have failed to confirm it. Many of the putative outlying foresights have proved wrong or are subject to controversy; most of the six suggested for Stonehenge, for instance, have turned out to be of historic rather than prehistoric origin. There are other examples. At best the jury still seems to be out on most issues.

But the view is not all negative. As Clive Ruggles, a worker in the field, puts it: "In many archaeoastronomical papers today one sees attempts to consider the astronomical and statistical evidence alongside, and on equal terms with, the anthropological and the ethnohistoric. There is far to go: but in drawing attention to problem areas where such collaboration is necessary, Thom's work may prove in the longer term to have opened up the interdisciplinary arena for a fascinating exchange of views across the 'two cultures.' . . . This may well prove to be the most significant benefit of all to be derived from the work of Alexander Thom."

The Shape of the Earth. I

I sometimes think how remarkable it is to have lived in the generation that first set foot upon a body beyond the earth. The first moon walk, more than twenty years ago, has made this a unique epoch in human history. Yet the *Apollo 11* mission to the moon had a total flight time of only 195 hours and 18 minutes. It was watched on television by an enormous audience—more than seven hundred million people—around the world, and passed from beginning to end with scarcely a hitch. Compare that excursion in July of 1969 with the campaigns sent by the French Academy of Sciences in 1735 to determine the shape of the earth. The first French expedition, dispatched to South America, was not heard from for nearly ten years. During their passage, the crew members suffered murder, mayhem, and incredible hardships. The public, so far as we know, cared not a jot.

The story of the expedition to measure the earth begins, interestingly enough, with the measurement of time. In 1672 a French scholar, Jean Richer, was sent to Cayenne in the colony of French Guiana in order to make some astronomical observations. Despite having carefully adjusted his pendulum clock in Paris before leaving, Richer discovered that the clock lost two and a half minutes every day when it was set up in Cayenne. Richer's observations remained a mystery until Isaac Newton seized on the time discrepancy to conclude that the earth must be an oblate spheroid, bulging at the equator. Newton argued that Cayenne, at five degrees north latitude, must be farther from the center of the earth than Paris at forty-nine degrees north latitude, and so experiences less gravity. The reduction in the earth's gravitational pull at the equator would cause a pendulum clock to have a longer period.

First published in *American Scientist*, 79:108, March–April 1991.

Newton's hypothesis could be checked by measuring an arc on the earth's surface corresponding to one degree of the earth's circumference; on a true sphere the arc length is everywhere the same, but on an oblate spheroid the arc increases with increasing latitude, as illustrated in figure 7.1. This test was carried out by a series of measurements in various parts of France in the early 1700s, with the triumphant result that the English were yet again in the wrong: the earth was in fact prolate—stretched out toward the poles! As so often in eighteenth-century English-French discussions, this one became rather heated, and eventually the French Academy of Sciences decided that the most sensitive and therefore the most conclusive test would be to measure two arcs, one as near the equator as possible, the other as near the North Pole as possible.

The equatorial expedition set out in April 1735, nominally led by Louis Godin, a senior member of the French Academy of Sciences. In practice, however, the expedition was led by Pierre Bouguer and especially by Charles-Marie de La Condamine, who was practical to the point of starting up silk and Holland lace boutiques en route. In fact, historians often refer to the voyage as La Condamine's expedition. The three men, a handful of colleagues, and their crew were destined for Ecuador (then a part of Peru) on the Pacific coast of South America, where they were to do their survey.

Progress was leisurely. In Cuba the expedition waited three months for permission from the local authorities to proceed; in Cartagena, Colombia, they waited four months. These delays, however, gave them time to contemplate alternative routes. One possibility was to travel southward from Cartagena up the Magdalena River for four hundred miles, and then on foot and mule for another five hundred miles over the Andes. They decided on the second route: by ship to Portobelo on the northern coast of Panama, across the isthmus, and then south to Guayaquil on the Ecuadoran coast.

In Portobelo the expedition was yet again delayed, but this time the wait was not at all welcome. The oppressive heat, heavy rains, and endless varieties of fever meant, according to one report, "that a large percentage of any ship's crew arriving there were buried before the ship was able to depart." From here they paddled and poled their way in dugouts against the fast current of the Chagres River, noted for its alligators, with a final trek on foot to the Pacific coast. Another

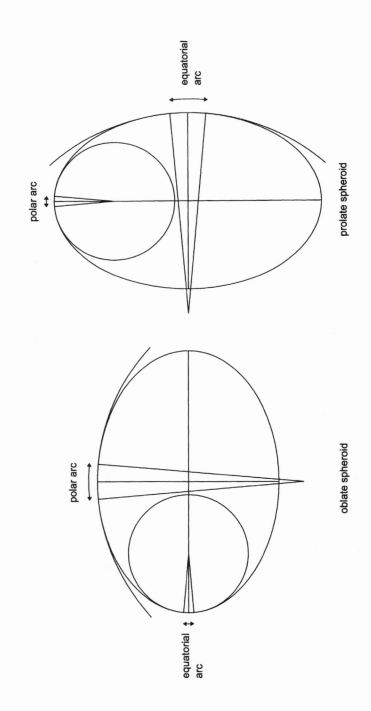

eight hundred miles south, and they were on the coast of Ecuador just south of the equator. Nearly a year had passed since they left Paris.

La Condamine had hoped that the coastal region would prove suitable for the measurement of the arc, but again their hopes were unrewarded. Not only was the coastal region unsuitable for survey work; it was completely inimical to the European visitor. The heat, a searing 1,028 degrees Réaumur (95 degrees Fahrenheit), was accompanied by intense humidity; snakes and scorpions abounded, and the insects swarmed so densely that they would extinguish a candle in minutes. Although the men took shelter behind special clothes and nets, the mosquitoes left them with swollen faces and enormous bites.

The expedition turned eastward and climbed upward into the Andes, in search of a high plain where their work could be done. They trudged into the jungle under an impenetrable canopy, through a watery morass that rose up to the horses' knees. As the elevation increased, they were forced to navigate precipitous river gorges and cross bridges even more dangerous than the rapids below. By May 1736, they were crossing the Chimborazo Desert and beginning to feel the altitude, wind, and now freezing temperatures. As they climbed the lower slopes of the 20,600-foot Chimborazo volcano, their pace was reduced to only a few miles a day, as their heavily laden mules struggled for a footing and needed to rest every seven or eight steps.

It must have been with no little thanks that the party finally sighted the capital city of Quito, where the Jesuit fathers made them welcome. Even here, at 9,400 feet, "all were at first considerably incommoded by the rarefaction of the air, particularly those who had delicate lungs . . . and were subject to little hemorrhages."

Figure 7.1 (*opposite*) *One way to decide whether the earth is oblate (left diagram) or prolate (right diagram) is to compare the distance along the earth's surface corresponding to a degree of geographic latitude near the equator with the distance corresponding to a degree near the poles. Geographic latitude is defined by the local curvature of the earth's surface. On an oblate spheroid, the radius of curvature is greater near the poles, and so the distance corresponding to a degree is also greater there than near the equator. The opposite is true on a prolate spheroid. (Original diagram by Elyse Carter,* American Scientist.*)*

At last, however, there were plains between the mountain ranges that would be suitable for their task. But no sooner had they started laying out a baseline than one of the group died suddenly. Monsieur Couplet's "distemper rose to such a height that he had only two days to prepare for his passage to eternity; but we had the satisfaction to see he performed his part with exemplary devotion. This almost subitaneous death of a person in the flower of his age was the most alarming, as none of us could discover the nature of his disease." Couplet's death was followed by that of an Indian helper, and Godin himself was frequently ill.

Despite such setbacks, the seven-mile baseline was measured forward and backward with agreement to within three inches. In the measurement of their arc, the surveyors planned to triangulate southward for over two hundred miles (almost three degrees of latitude), occupying nearly seventy survey stations. But before this undertaking could begin, they ran out of money. It was now January 1737, and La Condamine set off across the mountains again on a journey of fifteen hundred miles to Lima to arrange for new funding.

During La Condamine's absence, a new president was installed in Quito, and this somehow resulted in the party being accused of illicit trading. Unable to make any headway with the new authorities, one of the senior members hastened after La Condamine to summon help, while the accused took refuge in the Jesuit college. La Condamine returned, several months later, with new funding and documents proving that all traded materials had been correctly imported and that no one had done anything illicit. The survey could continue.

Triangulation began with one of the most difficult survey stations, on Pic de Pitchincha at over fourteen thousand feet. The summit was swept by relentless icy winds, and the sightlines were only briefly visible as cloud, fog, snow, and hail enveloped the workers—their faces, feet, and hands swollen from chilblains and painful almost beyond endurance. Breathing was an effort. For three weeks they huddled in a tiny hut that they had hauled piece by piece up the mountainside. They were sustained only by rations of boiled rice, fowl, and melted snow as they hovered over a chafing dish of coals. In all it required more than three months to complete the measurements at this one station.

As the work proceeded and the triangulation extended southward, the surveyors were surprised that they were unable to observe their previously occupied stations. When they returned to the sites, they found that the station signals had vanished. The signals had been constructed of wood and rope, which the local people decided could be put to better use, especially since almost no one had the slightest conception of what the foreigners were really doing. The best bet seemed to be that they were illegal prospectors hiding their true purpose behind some incomprehensible mumbo-jumbo story. The signals had to be reconstructed out of less desirable materials.

And so the survey continued, month by month and then year by year. Past beautiful but deadly Cotapaxi, a nineteen-thousand-foot volcano that could bring death to thousands, even in that sparsely populated land. Then Sangay, a volcano that had been erupting continuously for ten years with such violence that its explosions could be heard in Quito, 170 miles away. Particularly difficult were the frequent river crossings, the fast-flowing water having cut deep, narrow canyons that were bridged only by oxhide ropes either in the form of a *bujucos* (a primitive slatted bridge) or a *tarabita* (a bosun's chair arrangement). In either case, the mules had to be taken down to the river and made to swim.

In addition to performing the triangulation, the surveyors had to maintain and calibrate their instruments, as well as make astronomical observations to determine the latitudes at the ends of the arc. All in all, it was nearly 1740 by the time the observations were complete.

And then, back in Quito, the expedition's surgeon, Dr. Jean Senièrgues, was murdered. The doctor had been treating the daughter of a well-established local family, and gossip about their relationship began to spread, apparently fueled by the young woman's ex-fiancé. The furious doctor challenged the man to a duel, but in some way not quite clear, the surgeon was set upon by a mob that was leaving a bullfight. Although Bouguer, La Condamine, and others sprang to defend him, Senièrgues was fatally wounded. Outraged, La Condamine demanded justice, but it would be three years before the guilty were tried and, as a sentence, banished from Quito.

For Bouguer and La Condamine as well as a few other senior members of the expedition, it was time to go home. It is not clear what became of many of the lesser members; each seems to have proceeded as he saw fit, and some eventually straggled back to France from the

most unlikely corners of the globe. Although charged with writing
the official report, Louis Godin decided to take a post as an astronomer at the University of San Marcos in Lima. His cousin Jean Godin
decided to stay in Ecuador and marry a local thirteen-year-old girl.
Bouguer and La Condamine left Quito on separate routes in early
1743. Bouguer traveled north via the Magdalena River and returned
to Cartagena on the Caribbean coast. La Condamine was to do no
less than descend the eastern slopes of the Andes and travel the length
of the Amazon to the Atlantic. The precious records of their observations went with them; neither knew for sure what the implications for
the shape of the earth would prove to be.

The Shape of the Earth. II

In March of 1743, the French expedition in Ecuador had finished its measurements of the earth's surface. Eight long years had passed since Louis Godin, Pierre Bouguer, Charles-Marie de La Condamine, and their assistants had left France to measure an arc of latitude at the equator. Their results were to establish whether the earth is an oblate or a prolate spheroid. The expedition had trudged through the jungle, scaled the volcanoes and the frozen pinnacles of the Andes, and suffered deaths by disease, misadventure, madness, and even murder. Now it was time to go home, to compare their scientific records with those of a similar expedition sent to Lapland.

As noted in the preceding chapter, various members of the expedition chose to return by different routes. Godin, the nominal leader, first accepted the post of astronomer at the University of San Marcos in Lima. Bouguer chose to return by a relatively safe northern route through Cartagena on the Caribbean coast. La Condamine, on the other hand, chose a more adventurous route—down the Amazon River. Sparked by a Jesuit map of the Amazon (and the knowledge that no Frenchman had ever traveled it), La Condamine decided to descend the eastern slopes of the Andes and journey the great river's three-thousand-mile length to the Atlantic. He might have given thought to the words of the Spanish explorer Francisco de Orellana, who had preceded him by two hundred years: "Having eaten our shoes and saddles boiled with a few herbs, we set out to reach the Kingdom of Gold."

La Condamine left Tarqui on 11 May 1743, traveling south toward Lima over a difficult route through Zaruma, Ecuador. There was an easier path, but word had it that assassins were awaiting him on the

First published in *American Scientist*, 79:393, September–October 1991.

main route—intent on retaliation for his role in defending the murdered Dr. Senièrgues. It was slow trekking through the forests on the
lower mountain slopes. Rain fell for days, and the rivers were too
rapid and broken for rafting. At one river crossing, a mule, fully laden
with La Condamine's instruments and records, broke loose and fell
into the water. Drying the paper sheets, one by one, took several
painstaking days.

Down to the foothills and striking eastward into the Amazon basin,
the rivers became more navigable, though still treacherous. At one
point along the river Pongo, La Condamine's raft was carried along
at some fifteen feet per second. In the past thirty-six hours the level
of the river had fallen twenty-five feet—previously submerged tree
trunks now thrust through the water. During the night a half-
submerged branch speared the raft. Had La Condamine not been
awake, he would have been tossed into the rapid currents of the river.
At last, the Pongo broke with a roar through a defile and spread across
the wide plain, giving birth to the Rio Amazonas.

La Condamine was delighted. "I found myself in a new world, separated from all human intercourse, on a freshwater sea . . . penetrating
in every direction the gloom of an immense forest. . . . New plants,
new animals, and new races of men were exhibited to my view. Ac-
customed during seven years to mountains lost in clouds, I was rapt
in admiration."

And so down the seemingly endless Amazon. The indefatigable La
Condamine experimented at every turn. He noted that the natives
were hunting with blowguns, using darts dipped in the black resinous
poison curare. Would it be safe to eat game killed by poison? "There
is no danger," he reports cheerfully. "The venom of this poison is
only mortal when absorbed by the blood. The antidote is salt, but of
safe dependence—sugar." He tells of how he shot a chicken with a
curare-dipped arrow, pulled out the arrow, administered sugar, and
the chicken "exhibited no sign of the least inconvenience."

La Condamine was intrigued by the rubber trees he found everywhere along the banks of the Amazon. He had first encountered the
trees near the Pacific coast, but now he had time to marvel at the way
rubber "when fresh could, by means of molds, take any shape given
to it, at pleasure." He collected some raw rubber and a few utensils

made of the novel substance by the local people, for introduction to France.

Still eight days' sailing from the coast, the river was like a sea; its immense breadth heaved with the ebb and flow of the Atlantic tide. In late September, after four months of travel through mountain and jungle, La Condamine found himself welcomed in the coastal city of Pará (now called Belém). A scientist to the core, he immediately checked the latitude and longitude, as well as the length of the seconds pendulum. He needed to reach Cayenne in French Guiana, some five hundred miles to the north. He again chose the more difficult route. An ocean-going boat would have taken about two weeks. Instead, he elected to travel along the coast—in a twenty-two-oar dugout canoe—so that he might study the many mouths of the Amazon. The canoe trip took two months.

When La Condamine reached Cayenne, he had to wait six more months for a ship to France. He spent the time studying sea cows and quinine trees and experimenting on the speed of sound. And he slowly came to realize that his journey to the Pacific, into the Andes, and down the Amazon had been unnecessary. Contrary to preexpedition reports, French Guiana itself offered suitable terrain and a four-degree arc of equatorial latitude. The measurements could have been made without ever leaving French territory! One suspects, however, that La Condamine regretted little; he was a Renaissance man who met challenge with enthusiasm.

La Condamine left Cayenne in August of 1744, convalescing from jaundice, and after a stormy crossing of the Atlantic and two brushes with British cruisers found himself in Amsterdam at the end of November. After two months of waiting to obtain a passport to cross the Low Countries, he once more set foot in Paris on 23 February 1745. He had been gone ten years.

Several other members of the expedition had equally remarkable adventures. The return of Antonio de Ulloa, a Spanish naval officer who was attached to the expedition for political reasons, makes an especially pleasing story. Ulloa started his return from Ecuador in 1744 on a French ship that struggled around Cape Horn on a five-month voyage into the mid-Atlantic. The ship found itself in a cat-and-mouse game with British men-of-war, which eventually captured their prey. After a spell in a Newfoundland prison, Ulloa was sent to England, where he was incarcerated in Porchester Castle. When his expeditionary records were confiscated, Ulloa pleaded with his captors not to lose or destroy them. The appeal reached Martin Folkes, president of the Royal Society, who insisted the papers be sent to him so that they would not fall into the hands of the "ignoramuses" at the British Admiralty. Greatly impressed by the papers, Folkes arranged for Ulloa's release and had him quickly elected to the Royal Society! Ulloa returned in triumph to Madrid in 1746.

Nothing more bespeaks the character of the age, however, than the story of Jean Godin (the cousin of Louis Godin, the expedition's leader) and his wife, Doña Isabela. The courage of this indomitable woman would become the talk of Europe for decades.

Doña Isabela had been a thirteen-year-old girl when Jean Godin married her in 1743. The couple lived in Quito for some six years before they decided it was time to leave for France.

Godin elected to travel along the route chosen earlier by La Condamine—down the Amazon and on to Cayenne. He decided to test the route alone, then return to fetch his wife and children, and finally traverse mountains and river once more with the family en route to France. As Godin put it drily in a letter to La Condamine, "Travels in South America are undertaken with much less concern than in Europe; and those I had made . . . had made me perfectly a veteran."

In Cayenne, alone at the end of the first leg of the journey, Godin sought to obtain permission from the Portuguese to return up the Amazon. But the authority to travel again was slow in coming; several mishaps brought incredible delays. Godin began to suffer delusions, with symptoms suggesting paranoia. He saw the Portuguese as his enemy and sent wild plans to Paris for a French takeover of the Amazon basin. He built his own vessel and set sail for the Amazon, but immediately turned back at the thought of being captured by the Portuguese. In reality the Portuguese had sent a ship to fetch him, complete with an escort and instructions to take him up the Amazon, where they were to wait for his return with his family. Godin refused to set foot on board. The embarrassed French governor of Cayenne exchanged abusive letters with Godin. Eventually a messenger was sent in Godin's place, carrying a packet of letters for Doña Isabela.

Sixteen years had passed. Doña Isabela never received the letters, but word of their existence had reached her. Finally she decided she must set out herself to reach Godin. All four of their children had died of tropical diseases. She sent a slave, Joachim, to the mid-reaches of the Amazon to check the rumors that a ship awaited her there. Joachim returned in two years to say that indeed it was there—patiently waiting to take her to her husband. And so in 1769, Doña Isabela, now forty years old, set out at the head of a small party on a journey across the cordilleras and down the Amazon (variously called the dark river of tragedy and the living green hell).

Where they had expected to find guides, they found none. An epidemic of smallpox had ravaged the main routes, and the local populace had fled the area. The party rafted on down the wild rivers, hemmed in by the jungle walls, going at the pace of a cantering horse, slamming into rocks and submerged trees. Some food supplies and several of the men were lost overboard. Mosquitoes, black *piume* flies, and blood-sucking *jejenes* swarmed over their unprotected bodies until the blood flowed and people were driven half-mad.

It was decided that Doña Isabela, with three women servants and three men, would make camp, while Joachim and two others would go down the river to seek help. Those who stayed behind had difficulty finding food, and their morale ebbed. One of the men began an orgy of assault and rape on the servants; later he developed a morbid horror of the dark and awoke screaming one night believing that a vampire bat was sucking blood from his toe. The menacing jungle seemed constantly to threaten death. They were all close to breaking.

A month passed without sign of Joachim; Doña Isabela decided he must have perished. She decreed that the rest of them must build a raft and push on. Their inexperience augured failure; hardly had they set sail when their raft was smashed against a half-submerged tree, and their remaining possessions were swept away by the river. The survivors struggled up the river bank, some so delirious that they walked aimlessly into the jungle and never returned. Doña Isabela awoke the next day to find she was the sole survivor. It was, she later wrote, the thought of her husband (from whom she had been separated for twenty years) waiting in Cayenne that drove her after two days to leave the rotting corpses of her erstwhile companions. She appropriated their shoes and set out, as if in a trance, through the jungle. Soon after she left, Joachim returned. Finding some of the party missing and the rest dead, he concluded that none had survived. His report that Doña Isabela had died reached Ecuador, Cayenne, and even Paris before the truth was known.

Doña Isabela plunged on for nine days, living off roots and palm cabbage, her clothes in tatters. She stumbled on a group of Shimigai Indians, who took her by canoe to a mission at Andoas in Peru. Father Romero, the mission chief, proposed returning her to her family in Ecuador. "I am surprised at your proposals, Padre," replied Madame Godin indignantly. "God has preserved me . . . in my wish to rejoin my husband." She set off in another canoe, at last to meet the Portuguese ship still patiently awaiting her. It was now 1770; she still had two thousand miles of the Amazon to travel.

Eventually, as she and the Portuguese escort sailed into Cayenne, Doña Isabela and Godin were rejoined. "On board this vessel," wrote Godin, "after twenty years' absence and a long endurance on either side of alarms and misfortunes, I again met with a cherished wife whom I had almost given over every hope of seeing again."

Recovery was slow, and it was not until 1773 that they took ship for France. Thirty-eight years after he had set out to help determine the shape of the earth, Jean Godin returned to Paris. La Condamine was there to greet him on the dock.

In the next chapter, we shall look at the Lapland expedition, and then see how all these monumental efforts worked out.

The Shape of the Earth. III

In the previous two chapters, I have been considering some of the adventures arising from the eighteenth-century controversy over the shape of the earth. On the one hand, it had been found that a pendulum clock, carefully regulated in Paris, lost time when it was set up at a site near the equator. From this Isaac Newton deduced that the latter site was farther from the earth's center of gravity, and that the earth therefore must be an oblate spheroid. On the other hand, measurements of the north-south distance corresponding to a one-degree change of latitude, which varied between southern and northern France, led Jacques Cassini, the director of the Paris Observatory, to conclude the opposite: that the earth is prolate.

Because the north-south extent of France is rather limited, there was some doubt—though not shared by Cassini—as to the weight of Cassini's conclusion. To settle the issue, as we have seen, the French Academy of Sciences decided to send two expeditions, each to measure a one-degree arc, one at the equator and the other at the Arctic Circle, a range of latitudes that must surely bring a definitive result.

These expeditions, setting out in the 1730s to Ecuador and Lapland, were among the earliest modern scientific expeditions. In previous chapters I described the Ecuador (sometimes called the Peru) expedition—led by Godin, Bouguer, and La Condamine—and the legendary hardships endured, from the windswept heights of the Andes to the Amazon jungles, from misunderstandings to murder. Here I wrap up the story with a look at the northern expedition.

Most of the party bound for Lapland left Paris in April 1736, a year after their colleagues had set out for South America. The leader of the party was Pierre Louis Moreau de Maupertuis, then in his late

First published in *American Scientist*, 80:125, March–April 1992.

thirties and a man of considerable talent and stature. A scholar of philosophy, music, and mathematics, he had given up a career in the French army to travel and work in science. Already a member of the French Academy of Sciences and the Royal Society of London, he was later elected to the Académie française and eventually became president of the Berlin Academy of Sciences. Accompanying Maupertuis were several other famous figures: Alexis-Claude Clairaut, an outstanding mathematician and—very useful in the field—mental-arithmetic prodigy; Anders Celsius, a Swedish astronomy professor whose interests in thermometry apparently developed considerably during his winter in Lapland; and Pierre-Charles Le Monnier, later royal astronomer to Louis XV. (Among Le Monnier's subsequent achievements were some ten observations of the planet Uranus prior to its discovery by Sir William Herschel. Le Monnier mistook it for a star, on one occasion recording the coordinates on a paper bag used for keeping wig powder, which was then carefully filed in the observatory library. These observations, like other prediscovery ones, were later useful in determining Uranus's orbit.)

The total party of some fifteen men, including assistants and servants, had as its destination Torneå (now Tornio), on the northernmost coast of the Baltic Sea in the Gulf of Bothnia. From Paris the party traveled to Dunkirk, where a carefully selected cook was added to the roster, and all then boarded ship (amid mutual congratulations on the calm sea) for their 1,800-mile voyage to the north.

Perhaps it was a voyage typical of the age. The first setback was the discovery that the passengers' sleeping quarters consisted of a communal space some three feet high between decks, into which one crawled to endure extreme stuffiness and stale air through the night. The second setback was the rapid deterioration of the weather; within a day or two, squalls caused the sea to be "very much swelled," and the passengers became hideously seasick as they and their baggage were flung mercilessly around. The view from deck brought sight of a Norwegian ship "of bad appearance" pursuing them; fortunately they were able to outrun it eventually.

Three weeks out brought them to Stockholm, where, no doubt somewhat pale and shaky, they dined with the king and queen of Sweden. This had been carefully arranged by the French ambassador, since the expedition was headed for what was then Swedish territory and all diplomatic courtesies had to be observed. For the same reason, Celsius's appointment to the French expedition was as much a political as a scientific decision.

Those with sufficient authority now opted to continue the journey by land rather than by ship, and two coach loads of them set off northward along the coastal road for the remaining seven hundred or so miles, while the rest pressed on haplessly by ship through the Gulf of Bothnia. Those in the coaches soon found their new mode of travel not without travail. The roadway became less and less passable the farther north they went, and the only available fresh horses proved to be untrained and frequently unwilling to pull coaches. The coaches, in turn, could be gotten across wide, deep rivers only by putting the front wheels in one hired boat, the back wheels in another, and then having all the erstwhile passengers row furiously until the veins stood out on their brows.

Some three weeks of this finally brought the full expedition together again in Torneå, at nearly sixty-six degrees north latitude. It was the summer solstice of 1736, and all of the explorers were rewarded by the spectacle of the midnight sun.

Next was the need to reconnoiter the area to find where best to lay out a baseline and survey a north-south arc. The land north of Torneå was not particularly suitable, and so a party was sent to explore the eastern coast of the gulf. The leader of the group recorded that they "took seven men who were to row and steer the boat, which was a common one, in which we embarked with two servants and provisions for a fortnight; that is to say, biscuit and some bottles of wine remaining of the stock laid in at Dunkirk." This side expedition, whether for gastronomic or other reasons, found little to approve of on the eastern coastline, so attention reverted to the Torneå region. The town lies near the mouth of the river of the same name (today the border between Sweden and Finland), and the river runs in a roughly southerly direction between tall mountains. It was decided to use the river for transport and to set beacons on appropriate mountains as survey stations for the arc.

On July 6 the scientists set off upriver, transported in seven boats by a company of Swedish soldiers. The going was incredibly difficult and averaged about one mile per hour. The boats were designed for the river and its violent cataracts, being made of thin fir boards held together with thread from reindeer sinews to maintain flexibility. Where sails were lacking, the soldiers simply cut and held a small fir tree, complete with branches, in place. Even so, the scientists and their equipment were off-loaded at most difficult points, and they had to clamber along the steep, treacherous banks while the boats attempted the cataracts. The equipment, although delicate and capable of considerable precision in measuring angles, was also massive; one sector required a team of six soldiers to carry it. At intervals the company had to scale a neighboring mountain, clear a sightline, establish a survey station, make painstaking observations of previous stations (often waiting days for fog to clear), and then again descend to the river. On one such mountain Maupertuis fell off a wet, slippery rock, and it was feared that he had broken a leg, but fortunately he was able to walk again within a few days.

The local populace, while friendly, was extremely poor and able to supply very little food. The people spoke only Finnish, which made for further difficulties. (In one village the visitors were relieved to be able to converse with the local priest in Latin.)

The worst curse was the insects. Everyone wore all-encompassing clothing and heavy gauze veils, but whenever the veil touched the skin, the insects were instantly there drawing blood. Meals had to be eaten sitting in thick smoke. Even so, "In order to eat our bread, for we had nothing else, we were obliged to be very quick in passing our hand under the veils which covered our faces; without this precaution we should . . . have swallowed as many [flies] as crumbs of bread." The hardened soldiers covered their faces with tar, but to little avail: "No sooner was a dish served but it was quite covered with [insects], while another swarm, with all the rapaciousness of birds of prey, was fluttering around to carry off some pieces."

The expedition slowly worked its way northward, crossed the Arctic Circle, and established its last station at Kittisvaara, near Pello. The work had taken months, and there was need to hasten back to Torneå on the coast. By late September, it was snowing and the temperature was fourteen degrees Fahrenheit. Another month and they reached Torneå, just as the river froze over.

That the expedition would spend the winter in Torneå was beyond question; what was debatable was whether they should attempt now to measure the baseline needed to convert the angles measured along the arc into distances, or wait until spring. Astonishingly, given the extreme conditions and very little daylight, they chose to do it in December.

The snow was two feet deep when they started to lay out their 8.4-mile baseline, which they hoped to measure with wooden rods to an accuracy of a few inches. They tied logs into a triangle pulled by horses as a sort of snowplow, but soon gave up and laid out the line over the snow. Two separate parties moved up and down the line; comparison of their independent measurements eventually established a precision of four inches. Le Monnier, out on the line during the holidays, foolishly sipped brandy from a silver goblet and found his tongue frozen to it. The spirit-of-wine thermometers froze at minus thirty-three degrees Fahrenheit. By January bottles of brandy were freezing in unheated rooms. The team went out to finish the angle measurements from the ends of the baseline to the survey stations. The mercury thermometers reported a temperature of minus fifty-one degrees Fahrenheit.

While Clairaut worked assiduously at reducing the data, the others carried out pendulum experiments, observed a lunar eclipse, improved the town plan, or simply read books. Winter gave way to spring, and spring to early summer. On 9 June 1737, it was time to go home.

As before, the expedition divided into land travelers and sea travelers, with agreement to meet in Stockholm. The usual difficulties en route prevailed, in addition to which the maritime party, perennial losers, found themselves in a heavy storm taking on water at such a rate that bailing buckets and pumps proved inadequate and the ship was saved only by the pilot running it aground. Luggage and equipment were salvaged with considerable effort. But perseverance must have been the watchword, for on 21 August the expedition concluded its work by visiting Versailles and giving an account of its adventures to Louis XV.

Although the Lapland expedition had left a year after the South American one, the northern explorers had been gone only eighteen months, which meant that they were back with their results years before anyone returned from Ecuador. (As we have seen, La Conda-mine took ten years to return, while Jean Godin was gone thirty-eight years before setting foot in Paris again.) From this one might surmise that years would pass before the original question about the shape of the earth could be answered. Not so. The sending of two expeditions to such extreme locations proved to have been excessive. Each party had taken with it the knowledge of the length of a one-degree arc in France, and, such were the extreme latitudes of the new measure-ments, it became obvious even from preliminary analysis in the field that the earth must be oblate. The length of a one-degree arc was found to be 56,734 toises (a toise being six feet) in Ecuador, 57,060 toises in Paris, and 57,422 toises in Lapland. In fact, detailed consid-erations later led to the suspicion that the value for the Lapland arc had been overestimated, and remeasurements in the nineteenth and twentieth centuries showed that indeed Maupertuis's arc, through an unfortunate combination of errors, all in one direction, was some two hundred toises too long. But meanwhile, in 1791 the French National Assembly had replaced the toise with the meter, intended to be one ten-millionth of the distance from the equator to the pole along the Paris meridian. The errors in the various arc measurements (we now know) resulted in the practical calibration of the meter be-ing some 0.02 percent short of the intended length.

Newton had died in 1727 and so did not hear the result, although it is difficult to believe he died with much sense of doubt on the mat-ter. But the generous Maupertuis, in a letter to James Bradley, third astronomer royal of Great Britain, wrote, "Thus, Sir, You See the Earth is Oblate, according to the Actual Measurements, as it has been already found by the Law of Staticks: and this flatness appears even more considerable than Sir Isaac Newton thought it" (albeit the "more considerable" was largely due to the faulty Lapland arc).

Not that Maupertuis was universally regarded as any kind of hero, especially in France. Voltaire belittled the Lapland expedition and spoke of Maupertuis as the flattener of the earth, while the Cassini clan derided him as Sir Isaac Maupertuis and not only doubted the accuracy of the Lapland work but cast aspersions on the integrity of the expedition's members. It would be 1744 before Cassini gave in and accepted an oblate earth. Perhaps these slanders were partly re-sponsible for Maupertuis's move to Germany, even though his first visit might have cost him his life. An interested spectator at the battle of Mollwitz between Prussian and Austrian forces, he suddenly found himself violently taken prisoner and was lucky to be released by an officer who recognized him. Maupertuis married a German lady and accepted the offer of Frederick the Great to become president of the Berlin Academy, but thereafter things went downhill. His presidency was marred by controversy, and tuberculosis eventually led to his death at the age of sixty while visiting his friend Johann Bernoulli in Basel.

Celsius died at the even earlier age of forty-two without having achieved anything memorable beyond his work in Lapland, where he had been principal surveyor. Our present Celsius temperature scale is something of a misnomer; Celsius himself set the freezing point of water at one hundred degrees and the boiling point at zero degrees. It was Linnaeus who reversed the scale, but a later textbook attributed the modified scale to Celsius, and the name has remained.

Clairaut, one of the most precocious mathematicians of all time (he studied calculus at age ten and read a paper to the French Academy of Sciences at thirteen as author of the first treatise on solid analytic geometry), went on to ever greater fame in pure and applied mathematics. He became famous in astronomy for calculating (for the first time) the return date of Halley's comet in 1759, allowing for planetary perturbations. The accuracy of his prediction was a triumph not only for his own calculating powers but also for Newtonian mechanics.

Few among those of the equatorial expedition went on to other memorable achievements. Bouguer became a professor of hydrography and is remembered in astronomy for his pioneering work in photometry. He succeeded Maupertuis as chief geometer of the Academy of Sciences when Maupertuis left for Berlin.

Perhaps the most likable of them all, though, was La Condamine. A man of irrepressible spirit, he was a scientist through and through and never stopped inquiring about and experimenting with everything around him. Typically, as age and paralysis overtook him in his seventies, he offered a prize of a thousand francs to any physician who could diagnose the cause of his paralysis (not necessarily cure it), and while awaiting the decision of the Berlin Academy as to the winner (there was none) took up song writing.

What emerges strongly from accounts of these early scientific expeditions and others like them is the amazing fortitude of the men and women (remembering Doña Isabela) who undertook them. They faced incredible hardships and often risked their lives to a degree unthinkable today, reporting it all quite casually. One can only hope the spirit, if not the need, lives on.

The Last Universalist

It would be interesting, perhaps amusing, to survey the answers one might get to the following question: What mid-nineteenth-century British scientist was to his time what Einstein has been to our time (in terms of being a household name), and was so revered that he was buried close to Sir Isaac Newton in Westminster Abbey? I wonder how many would guess John Frederick William Herschel.

This year [1992] marks the two hundredth anniversary of John Herschel's birth. Although the occasion is being commemorated by various groups around the world, its impact on the scientific community is probably nil. If ever there was a person of whom it can be said that fame was fleeting, it was John Herschel.

John was the only child of William Herschel, a figure much more readily remembered in the astronomical community. An astronomer who speaks simply of "Herschel" refers to William. Not surprisingly, this father-and-son pair created interesting contrasts. John was the product of wealth and the finest education, a Cambridge man of outstanding intellect and social connections. A generation earlier, William had escaped life as a teenage band player in the Hanoverian army with only the proverbial penny to his name, and later turned up in England as an obscure musician. After that, William was an organist and music teacher in the town of Bath. In his spare time he began using and then building astronomical telescopes. He was aided in this work by his remarkable sister, Caroline, who also gave up a career in music, in her case to keep house for William. A woman deserving far greater recognition, Caroline became deeply involved in astronomy: she discovered comets and nebulae, and edited and revised

First published in *American Scientist*, 80:422, September–October 1992.

catalogues. These and other activities eventually earned her the gold medal of the Royal Astronomical Society.

In 1781, Caroline's and William's lives changed forever when he accidentally discovered Uranus—the first planet to be discovered with a telescope. The resulting fame brought William's appointment as astronomer to King George III, and stipends for himself and Caroline (arguably making her the first professional woman astronomer). William became a full-time astronomer, assisted at every turn by Caroline. He not only built professional-quality telescopes but also used the best of them to make discoveries that brought him recognition as the greatest observer of his age.

Although he made a fair profit from the sale of hundreds of telescopes, the work was not without hazard. Early experiments involved large metal mirrors that were cast in molds of horse dung. On one occasion a mold broke and molten metal poured across the floor, producing a precipitous race for the door on the part of William and his assistants.

At the age of fifty, William unexpectedly married a neighbor's widow. It was a bitter blow to Caroline. She moved away, no longer her brother's professional associate and helper.

John, the sole child from the marriage, arrived when William was fifty-three, an age difference that was perhaps difficult to bridge. The household was always quiet because William spent many daylight hours asleep, and young John was rescued from solitude and silence only by visiting his beloved aunt Caroline, a lively and joyful woman. Later, when Caroline moved back to Hanover after William's death, John never missed an opportunity to visit her. He described with delight a visit when Caroline was eighty-three: "She runs about the town with me, and skips up her two flights of stairs. In the morning until eleven or twelve she is dull and weary, but as the day advances she gains life, and is quite 'fresh and funny' at ten o'clock P.M., and sings old rhymes, nay even dances, to the great delight of all who see her." On her ninety-seventh birthday, Caroline sang a composition of her brother's for the prince and princess of Prussia. She lived to be ninety-eight.

John's early education came from a private tutor, and he later attended Eton College and Cambridge University. His abilities were quickly evident. With Charles Babbage and George Peacock, John formed the Analytical Society at Cambridge, which was devoted to bringing the faltering field of British mathematics up to the standards prevailing in Europe. John's mathematical gifts took him through the tripos exams as senior wrangler and to first Smith's prizeman. (Charles Babbage backed out because he felt unable to compete against Herschel.) When he was twenty-one, these same skills won John fellowship in the Royal Society, and soon thereafter he won the society's Copley Medal. Nevertheless, his first career move was to stand for the chair of chemistry at Cambridge, which he lost by one vote at the age of twenty-three. Declining his father's urging of a career in the church, John went to Lincoln's Inn to start a career in law.

Within a year, however, John abandoned all thoughts of law and returned to science. His father, then seventy-eight, was worried about his own declining powers, and his as-yet-incomplete research. The twenty-four-year-old John, with a sigh, it seems, took on a filial duty—completing his father's research. John was regarded as a shoo-in for the Lucasian Professorship of Mathematics at Cambridge (once held by Newton), and was offered the chair of mathematics at the University of London, but declined it. It says much of his character that, with multiple doors open before him, he took on a task requiring many, many nights of dusk-to-dawn work under bitter conditions, even though he had little taste for the task.

Nevertheless, it was not all work. In the early years, John made a number of journeys abroad. He and Babbage traveled through Europe, doing serious mountaineering in the Alps and also meeting Joseph Louis Gay-Lussac, Siméon Poisson, Jean-Baptiste-Joseph Fourier, Giuseppe Piazzi, Karl Friedrich Gauss, Friedrich Bessel, and others, most of whom John later kept in touch with through the half-dozen or so European languages in which he was fluent. His journeys, which earlier had included a meeting with Napoleon, later extended to Ireland, where he found a close mathematical colleague in William Rowan Hamilton. Between journeys John intermingled the production of observational catalogs of 2,307 nebulae and 3,346 double stars with mathematical research. This work brought him further medals from the Royal Society, the Lalande Prize of the French Academy, the presidency of the British Association for the Advancement of Science, and the foreign secretaryship and three times the presidency of the Royal Astronomical Society, as well as its gold medal. In his spare time he contributed long articles to encyclopedias.

Although still in his thirties, John had earned almost every possible distinction in his field. To round out his life, he married a charming, practical, remarkably beautiful woman, Margaret Brodie Stewart. They had twelve children, with whom John, perhaps remembering his solitary childhood, had a warm and happy relationship.

It seems to have been long on John's mind that, although he and his father made huge advances in cataloguing the nebulae and double stars of the Northern Hemisphere, a vast lacuna remained in the southern skies. Thus it was that in 1833 John and his family embarked for Cape Town, a gateway to what was then very much darkest Africa for European travelers. The British Admiralty had established an observatory there as the southern counterpart to Greenwich. Thomas Maclear, the latest director of the observatory, arrived only ten days before the Herschels. (After months of seasickness, Maclear was greeted on the dock by his chief assistant with the memorable words, "So, Sir, you have determined to accept this wretched appointment!") John, as a private individual, established himself on an estate some miles away, but his friendly relationship with Maclear led to pleasant times warmly recalled in his published diaries and correspondence.

Although others on John's voyage had been utterly prostrated by seasickness, he had occupied himself with dissecting sharks, detailing sea and weather conditions, and comparing navigation results with the ship's officers. Upon arriving in Cape Town, he enthusiastically set up telescopes with which he catalogued 1,707 nebulae and 2,102 double stars, made detailed sketches and inventories of the Magellanic Clouds, witnessed the only historical outburst of the strange object now known as η Carinae (a unique star in the Southern Hemisphere that has increased and decreased in visual intensity over time), devised a new method of astronomical photometry, observed eclipses, helped solve Maclear's problems with the observatory's instruments, and completed endless other astronomical tasks. John galvanized the Cape Philosophical Society, participated in exploratory expeditions, used a camera lucida to sketch Cape scenes, did enough botany to have his name appended to a number of species, measured tides and winds, drew maps, devised an education system for the colony that has lasted into our time, and on and on. And all the while his marvelous wife, "Maggie," pregnant for more than half her stay at the Cape, managed a household of seldom less than twenty, basking in the love of her adored husband.

The family's time in Africa came to an end in 1838. John wrote to his brother-in-law: "Whether much good . . . or much evil is in store for us on our return to Europe is uncertain . . . but this I know that we have been very happy here . . . and that our residence at the Cape . . . come what may will always be to me . . . the Sunny Spot in my whole life."

Back in England, John (now Sir John) settled down to middle age, already an elder statesman of his time. With his filial duty discharged, his observing days were largely over, and he returned to his earlier love of chemistry, mainly through research in photography. He discovered the action of "hypo"—a fixative, sodium thiosulfate, that renders photographic paper insensitive to light after developing, and he invented the terms "positive" and "negative" as they are used in photography. But commissions and committees increasingly occupied his time. In 1850, he became master of the mint, like Newton before him. Here he made major reforms and advocated the introduction of decimal coinage, an idea a century ahead of its time in Britain. And everywhere one turns in astronomy, one finds a touch of John Herschel's hand, from the definition of constellation boundaries to the great catalogues of nebulae that eventually became the *New General Catalogue*.

For relaxation John translated great works into English. He translated Friedrich Schiller's German verse into English hexameters. John also translated from Greek (Homer's *Iliad*) and Italian (Dante's *Inferno*).

By his mid-sixties, John was clearly a man in physical decline, although his mind was unimpaired. Suffering greatly from bronchitis and gout, he soldiered on into his seventies, grieving the passing of such old friends as William Whewell and Augustus De Morgan with the words, "Life has no glare to a man entering on his 75th year. It is the shade and the calm that he longs for." It was in shade and calm that John Herschel died in 1871, at the age of seventy-nine. He was a hero of his age, hailed as the last universalist.

What can we make of the fascinating contrasts between father and son, and the paradox of their legacies? William was, to give him his due, the foremost observational astronomer of the eighteenth century, even though he was never trained in astronomy or mathematics. Nonetheless, his discoveries were mainly serendipitous, and his theories were often ludicrously wrong even for his time. (For example, he claimed that the moon is inhabited and that the sun is a cool body, with mountains pushing through a fiery envelope to form sunspots.) Nevertheless, William's fame continues undiminished. John, however, was a man of enormous brilliance and had the best intellectual tools at his fingertips. Using these tools, he made fundamental contributions in many fields, and was rarely wrong. In his day, John was regarded as Einstein has been regarded in our day. Yet today, John is dismissed as a figure of little consequence.

In part, I suppose the cause of this asymmetry in fame is obvious. William discovered Uranus, earning headlines and popular-book entries that none of John's arcane mathematical proofs ever attracted. But there is more to it than that. At the age of thirty-four, John wrote of "a kind of obscure consciousness that I am not destined . . . to make giant inroads into great branches of human knowledge . . . but rather to loiter on the shores of the ocean of sciences and pick up such shells and pebbles as take my fancy for the pleasure of arranging them and seeing them look pretty." Arthur Eddington would later refer disparagingly to such activity as "stamp collecting." Personally, I find it a refreshing change from our day, when granting agencies place no weight on research for fun and pleasure in themselves. The tide of history may have washed over John Herschel with little more than a ripple, but we can concur with his biographer: "The warmth of his humanity, so richly enjoyed by his contemporaries, can still be appreciated in a colder and more practical age."

The Great Moon Hoax

In the previous chapter I discussed Sir John Herschel, a mid-nine-teenth-century figure who was one of the most famous scientists of his age. It was perhaps this fame that led to his being the subject of a marvelous hoax, known forever after as the Great Moon Hoax. Not that Herschel himself was in any way involved; in fact, the entire hoax was over before Herschel even heard of it. He was on a private expedition to South Africa when the hoax was perpetrated in a New York City newspaper in August 1835.

The perpetrator was a journalist, Richard Adams Locke, British-born and said to have been a descendant of the philosopher John Locke. His hoax took the form of a series of articles published in the *Sun*, purporting to be reprints from a supplement of the *Edinburgh Journal of Science* and describing the incredible discoveries suppos-edly being made by Herschel in South Africa with an equally incred-ible telescope.

Locke opened the series with a scene prior to Herschel's depar-ture from Britain. What kind of telescope should accompany him, mirror-based or lens-based? Herschel was an expert on the metallic mirrors of the day, which required a great deal of polishing; he cus-tomarily "watched their growing brightness under the hands of the artificer with more anxious hope than ever lover watched the eye of his mistress," wrote Locke. But when it came to studying insects on the moon, Herschel concluded he would need not a mirror but a lens, twenty-four feet in diameter and weighing 14,286 pounds. A lens introduced the problem of chromatic aberration (bringing light of different colors to the same focus), which was usually solved by making the primary lens a sandwich of two or more lenses having

First published in *American Scientist*, 81:120, March–April 1993.

different shapes and different densities of glass. It would be a tricky business matching components of this size, but Herschel in a moment of inspiration saw that simply melting all the lenses together "would as completely triumph over all impediments."

Such a stupendous lens would be thick and absorb light, making for a dim image. Even the redoubtable Herschel saw need to consult Sir David Brewster, an optical specialist, on this one. "The conversation became directed to that all-invincible enemy, the paucity of light in powerful magnifiers. After a few moments' silent thought, Sir John diffidently inquired whether it would not be possible to effect a *transfusion of artificial light through the focal object of vision!* Sir David sprung from his chair in an ecstasy of conviction, and leaping halfway to the ceiling, exclaimed, 'Thou art the man!'" The scheme was to illuminate the image with a "hydro-oxygen microscope," which would then project the image on a large canvas screen. A working model, made from "the shop window of Mons. Desanges, the jeweller to his ex-majesty Charles X, in High Street [Edinburgh]," was perfect.

The full-size 288-inch lens, however, would require "money, the wings of science as the sinews of war." Needless to say, though, the Royal Society was wildly enthusiastic about the project and proposed it "as a fit object for the privy purse. . . . His Majesty, on being informed that the estimated expense was £70,000, naively inquired if the costly instrument would conduce to any improvement in navigation? On being informed that it undoubtedly would, the sailor King promised a *carte blanche* for the amount which might be required." The giant lens was cast and cooled within a week and emerged "immaculately perfect."

Why now would Herschel make the arduous journey to South Africa with this enormous telescope? Locke pondered. The *Sun*'s readers, agog with expectation, could hardly wait. Locke delivered. Herschel had to go south because he also intended observing a transit of Mercury across the face of the sun, and regrettably this would take place when it was night in the Northern Hemisphere but daylight in the Southern Hemisphere!

So it was, the *Sun*'s readers were told, that Herschel, a Dr. Grant (the alleged author of the *Edinburgh Journal of Science* supplement), a "Lieutenant Drummond of the Royal Engineers, FRAS [Fellow of the Royal Astronomical Society], and a large party of the best English mechanics" set off for Cape Town. Having arrived after "an expeditious and agreeable passage," the group selected a site where, "aided by several companies of Dutch boors, [Herschel] proceeded at once to the erection of his gigantic fabric."

Locke skillfully wove references to real astronomy into his chimerical story. Halley's comet was due back in late 1835, an apparition that many people feared. It was already nearly September of that year, and superstitious fears were growing. Why not report that Herschel, with his wonderful telescope, had already sighted the approaching comet long before anyone else? The *Sun* did so, adding that this had put Herschel in a difficult position. The great telescope was supposed to be secret ("royal patrons [had] enjoined a masonic taciturnity"), but Herschel felt impelled to report his findings in the interests of calming the populace. Thus Locke reported that Herschel, without being able to say how he had come by his results, wrote

> to the astronomer-royal of Vienna, to inform him that the portentous comet predicted for the year 1835, which was to approach so near this trembling globe that we might hear the roaring of its fires, had turned upon another scent, and would not even shake a hair of its tail upon our hunting-grounds. At a loss to conceive by what extra authority he had made so bold a declaration, the men of science in Europe who were not acquainted with his secret, regarded his discovery with incredulous contumely, and continued to terrorize upon the strength of former predictions.

It is amusing to note that the real Herschel, experienced astronomer though he was, had considerable difficulty in locating Halley's comet from his observatory outside Cape Town. Working from a detailed, though evidently erroneous, ephemeris, he at first failed to find the comet, and only came on it belatedly. His triumphant announcement to his family brought the response that a junior assistant had already found it several nights earlier and shown it to Herschel's young son, James. "I never was more inclined to give a man hard words or even a hard knock," the irate Sir John confided to his diary.

The *Sun*'s readers, cometary fears perhaps allayed, were eager to get on to the real task of the telescope, which, it had been announced at the start, was a study of the moon. Locke kept them poised.

Herschel, after all, was "about to crown himself with a diadem of knowledge which would give him a conscious pre-eminence above every individual of his species. . . . He paused ere he broke the seal of the casket which contained it." Fortunately courage and Dr. Grant prevailed, and it

> was about half-past nine o'clock on the evening of January 10, the moon having then advanced within four days of her mean libration, that the astronomer adjusted his instruments for the inspection of her eastern limb. The whole immense power of his telescope was applied. The field of view was covered throughout its entire area with a vivid representation of basaltic rock. . . . This precipitous shelf was profusely covered with a dark red flower, "precisely similar" says Dr. Grant, "to the Papaver Rhoeas, or rose-poppy of our sublunary cornfield."
>
> [Later] a verdant declivity of great beauty appeared. They were delighted to perceive that novelty, a lunar forest. "The trees," says Dr. Grant, ". . . were of one unvaried kind, and unlike any I have seen, except the largest kind of yews in the English churchyards."

What of the lunar maria? Locke was sure his readers would prefer them to be real seas.

> We again slid in our magic lenses to survey the shores of the Mare Nubium. Fairer shores never angels coasted on a tour of pleasure. A beach of brilliant white sand, girt with wild castellated rocks, apparently of green marble . . . moved along our [screen] until we were speechless with admiration. The water, wherever we obtained a view of it, was nearly as blue as that of the deep ocean, and broke in large white billows upon the strand. [Jutting from the water was] a lofty chain of obelisk-shaped, or very slender pyramids, standing in irregular groups, each composed of about thirty or forty spires. Dr. Herschel shrewdly pronounced them quartz formations. . . . On introducing a lens, his conjecture was fully confirmed; they were monstrous amethysts, of a diluted claret color, glowing in the intensest light of the sun! They varied in height from sixty to ninety feet.

Moving on past a three-mile island of pure sapphire, "We frequently saw long lines of yellow metal hanging from the crevices of the horizontal strata in wild network, or straight pendant branches. We of course concluded that this was virgin gold."

The *Sun*'s readers were in an ecstasy rivaling that of Sir David

Brewster. Circulation, which prior to the hoax had been steady at eight thousand, was by now more than nineteen thousand, making the *Sun* temporarily one of the world's largest newspapers. Crowds besieged the paper's offices, clamoring for copies as fast as the presses could provide them. While waiting, the mob was assured by a highly respectable looking gentleman that he had actually seen the telescope being shipped in England, and that the whole thing was undoubtedly true.

The readers were beside themselves in anticipation of Herschel finding animal life on the moon. Locke was not one to deny them:

> Here our magnifiers blest our panting hopes with specimens of conscious existence. In the shade of the woods . . . we beheld continuous herds of brown quadrupeds having all the external characteristics of the bison. It had a remarkable fleshy appendage over the eyes, crossing the whole breadth of the forehead and united to the ears. This hairy veil was lifted and lowered by means of the ears. It immediately occurred to the acute mind of Dr. Herschel, that this was a providential contrivance to protect the eyes of the animal from the great extremes of light and darkness to which all the inhabitants of our side of the moon are periodically subjected.

A subsequent edition reported unicorns "of a bluish lead color," but the burning question was whether Herschel was going to find anything resembling human life. Sure enough,

> We were thrilled with astonishment to perceive four successive flocks of large winged creatures, wholly unlike any kind of birds, descend with a slow even motion from the cliffs on the western side, and alight upon the plain. They were first noticed by Dr. Herschel, who exclaimed, "Now gentlemen . . . we have here something worth looking at: I was confident that if ever we found beings in human shape, it would be in this longitude." Certainly they were like human beings, for their wings had now disappeared, and their attitude in walking was both erect and dignified. They averaged four feet in height, with short and glossy copper-colored hair, and had wings composed of a thin membrane, lying snugly on their backs [while] the face was a slight improvement upon that of the large orang outang. . . . Lieut. Drummond said they would look as well on a parade ground as some of the old cockney militia!

All this had occupied less than a week of the *Sun*'s editions, but

Locke must have been wondering how to wind it all down. The denouement came rather more suddenly than he had intended, when two Yale professors arrived and asked to see some of the "original" material for themselves, and a rival newspaper began to look into the story. Locke quickly confessed (to the triumphant rival paper) and soon parted company with the *Sun.*

The exposé of the hoax seems to have been met more with amusement than outrage. Edgar Allan Poe called Locke a genius, and noted that not one person in ten had suspected the story to be a hoax.

Reprints of the hoax soon abounded, and eventually a copy reached Herschel in South Africa. Even the usually excitable Sir John seems to have taken it in his stride. His only reference to it that we have is one sentence in a letter to his aunt Caroline: "I have been pestered from all quarters with that ridiculous hoax about the Moon—in English French Italian & German!!" His wife, Margaret, also writing to Caroline, called the story "a very clever piece of imagination" and said "the New Yorkists were not to be blamed for actually believing it."

As for Richard Locke, so far as I know this constituted his only fling at fame, and he passed into history solely as the author of an enduring hoax.

The Tunguska Event

At fourteen minutes and twenty-eight seconds past seven on the morning of 30 June 1908, S. B. Semenov was sitting on the open porch of his house in the village of Vanavara in central Siberia, when he witnessed what was perhaps the most brilliant flash of light ever seen in historical times. "My shirt almost burned off my back," he recalled. "I saw a huge fireball that covered an enormous part of the sky. I only had a moment to note the size of it. Afterward, it became dark and at the same time, I felt an explosion that threw me several feet from the porch. I lost consciousness . . . and when I came to, I heard a noise that shook the whole house and nearly moved it off its foundation. The glass and the framing of the house shattered."

At latitude 60 degrees, 55 minutes north, and longitude 101 degrees, 57 minutes east, the explosion site was one of the most remote places on earth, in an area north of the Podkamennaya [Stony] Tunguska River forty miles from Semenov's house. The phenomenon has been dubbed the Tunguska Event. It happened at an altitude of some twenty-eight thousand feet and was not only seen but also heard five hundred miles away. Peasants in the area reported "unbelievably loud and continuous thunder; the ground shook, burning trees fell, and all around there was smoke and haze." More than a thousand reindeer were killed, and several nomadic settlements vanished. Huge black clouds began to ascend miles into the stratosphere, later bringing a "black rain" of debris and dirt.

Nearly four hundred miles to the southwest, the Trans-Siberian express train was wildly jolted, and the engineer, incredulously sighting the vibrating rails ahead, brought the train to a screeching halt. Sounds of distant thunder followed.

First published in *American Scientist*, 81:412, September–October 1993.

Astonishing though it may seem today, more than a dozen years passed before anyone beyond central Siberia heard about any of this. Thousands of miles away in Germany, the United States, and Java, though, seismic detectors recorded the event, yielding its precise timing. Within five hours, meteorological stations across England recorded twenty minutes of wild fluctuations in barometric pressure. Magnetic disturbances were reported from London. In the following days, reports from northern Europe portrayed the night skies as so bright that photographs could be taken at midnight, and ships could be seen clearly for miles out to sea. By 4 July, the *Times* of London was noting the similarity of such events to those following the great volcanic explosion of Krakatoa in 1883, although "no volcanic outburst of abnormal violence has been reported lately."

No explanation of all this being at hand, the disturbances and seismic activity were dismissed vaguely as probably owing to "earth tremors," and they faded from the news. The least attention was paid in Russia, a country caught in unrest and social upheaval, sliding irrevocably on its century-long path to revolution. Only three years had passed since the army had massacred many in the crowd marching on the Winter Palace in St. Petersburg. Then came the First World War, followed by the Great Revolution of 1917. Civil war followed, and Siberia itself was wracked with horror and bloodshed as the Red and White armies fought ruthlessly and relentlessly across its endless spaces. "Siberia was thrown into anarchy," wrote Lengyel. "Typhus broke out among the refugees [and] they stampeded to the towns in the hope of finding relief. In one town of 70,000 inhabitants, struck by the refugee wave, about 200,000 perished. Thousands lay on the streets, at railway stations, along the roads. Tens of thousands of monstrously blown up human forms awaited merciful death in the town of Taiga alone." On the Yenisei River in the early winter of 1920, "Hundreds of bodies with heads and hands cut off, with mutilated faces and bodies half burned, with broken skulls, floated and mingled with the blocks of ice, looking for their graves."

Caught up in much of this was the young scientist who was to bring the attention of the world to the Tunguska Event. In 1921, Leonid Kulik was thirty-eight, a survivor of arrest and imprisonment for revolutionary activities, a veteran of three wars, and yet a "vibrant, cultured man." Now, working in the Mineralogical Museum of Petrograd (St. Petersburg), he had come to the study of meteorites.

Preparing for an expedition to Siberia to search for meteorite falls in general, Kulik came upon an old Siberian newspaper account of the great event of 1908. Soon he had unearthed others, and although they were vague and contradictory, he realized there must be some substance to them. They described an object moving, it seemed almost horizontally, through the early morning sky from a roughly southerly direction, accompanied by loud noise and earth tremors that culminated in the gigantic explosion near Vanavara. Unexpectedly, though, "The noise was considerable, but no stones fell." To Kulik's surprise, he found that no scientist had ever investigated the event. He resolved to do so.

So the first of many expeditions to study the event set out in 1921, with Kulik at its head. Some thirteen years had already passed; faced with accounts contradicting one another, Kulik had little to go on as to where the search should begin. He settled on the town of Kansk on the Trans-Siberian railroad, but although he found there many new tales and eyewitness reports from surrounding areas, he soon realized that he was still hundreds of miles from the epicenter of the burst itself.

Painstakingly, Kulik put together the many accounts of that terrible day in June 1908, but it was 1927 before he was able to mount a second expedition to find the site. The intervening work had revealed that the epicenter lay to the north of the Stony Tunguska River, but exploring the region would be a challenge. Even in more inhabited areas of Siberia, travel was difficult. Anton Chekhov had complained that "the Siberian highways have their scurvy little stations. . . . They pop up every 20 or 25 miles. You drive at night, on and on, until you feel giddy and ill, but keep on going, and if you venture to ask the driver how many miles it is to the next station, he invariably says, 'Not less than twelve.'"

In the more remote areas there were no roads at all. Trappers simply followed the tracks worn by the wildlife through what was described by Yuri Semyonov in a study of Siberia as a "vast and sinister" primeval forest in which "the weak and imprudent often perish." Distances were measured on two scales: summer and winter. Winter distances over the icy wastes were much shorter, even though it could become so cold that birds would drop, frozen, from the skies. In summer, one had to contend with pathless bogs where, as Semyonov puts it, "everything below is decayed and rotten, and everything above withered, where only the corpses of the huge trunks slowly molder away in the brackish water." An American prospector of the time reported, "I had a terrible journey through forests and over mountains, where rain fell incessantly and I nearly died through exhaustion. The valleys, hillsides, in fact everywhere we went, was covered with bog made by the falling pine-needles through countless ages. The horses plunged on for miles in this stuff up to their knees."

On 8 April 1927, Kulik and one assistant set out on pack horses along the banks of the Stony Tunguska. Five days later, exhausted and sick with scurvy and various infections from months of arduous travel and poor food, they crossed the Makirta River. There, standing on its banks, they finally stared in astonishment at the explosion site stretching to the horizon before them. "The results of even a cursory examination exceeded all the tales of the eyewitnesses and my wildest expectations," Kulik wrote.

Everywhere before them, the ancient forest lay smashed to the ground, the tree trunks in parallel ranks. "One has an uncanny feeling," wrote Kulik, "when one sees . . . 30-inch giant trees snapped across like twigs, and their tops hurled many meters away to the south." The parallelism and the fact that every dismembered branch showed signs of burning attested to the fact that this had been no mere forest fire; an unbelievably powerful flash had ignited them.

Kulik felt sure that not far beyond the horizon he would find some stupendous crater excavated in the moment of that flash. But though he crisscrossed the terrain through the following weeks, no crater appeared. Eventually, the south-pointing flattened trees were replaced by similar north-pointing ones, and Kulik came to realize he had traversed the blast area. Only a low swamp lay at its center, but equally unexpected was a forest of "telegraph poles," dead trees still standing, but with every branch and twig blown away.

Kulik would return again and again to investigate the Tunguska site, until in 1941, as a member of the Moscow People's Militia, the fifty-eight-year-old was caught up in his fourth war, was taken prisoner, and died of typhus in a camp near Smolensk on 24 April 1942.

In the postwar years, into the 1960s and beyond, new expeditions used ever more sophisticated means to investigate the Tunguska

Event. No conclusion was without controversy, although most Soviet investigators clung to the belief that it had been caused by a comet, or more likely a meteorite. New times brought new fashions, of course. When atomic energy, space travel, and the search for extraterrestrial life became topical, a popular theory saw the event as likely caused by an atomic-powered alien spacecraft spinning out of control. The notion of antimatter prompted so distinguished a scientist as Willard Libby to suggest with his collaborators in 1965 that the object had been composed of antimatter, an idea laid to rest by the 1969 Condon Report on UFOs, which cited the lack of evidence for radiocarbon production. Later, in 1973, when black holes were all the rage, it was suggested that a mini–black hole had struck Siberia, passing on through the earth and out the other side.

None of these exotic theories was widely accepted, and now it seems that three recent papers, returning to the original idea, may have brought speculation to a close.

Ten years ago Zdenek Sekanina, a scientist at the Jet Propulsion Laboratory in California, published an exhaustive study of the evidence surrounding the Tunguska Event. His analysis concluded that the object came in from a direction close to 110 degrees east of north at an angle about five degrees above horizontal, at a speed of some nineteen miles per second (about ten times the speed of a high-velocity rifle bullet), and exploded at an altitude of around 5.3 miles after resisting aerodynamic pressures exceeding a thousand times normal atmospheric pressure. Although eyewitness reports just prior to the explosion suggest a fireball fainter than the sun, the explosion itself was about forty times the brilliance of the noonday sun, implying a mass dissipation of a million tons of material in less than a tenth of a second, equivalent to the energy release of the first atomic bombs. This, of course, was preceded by the effects of the object's stupendous ballistic energy. All told, enough energy was released to wipe out even modern New York City.

The velocity vector, after allowing for acceleration by the earth's gravity, leads to a solar-system orbit that rules out a comet, as does the high resistance to aerodynamic pressures (comets being constituted mainly of ices and dust). Sekanina concluded that the object was most likely a small asteroid, some three hundred to six hundred feet across. Christopher Chyba, Paul Thomas, and Kevin Zahnle have reached conclusions quite similar to those of Sekanina, although they call for a steeper entrance angle and a somewhat smaller size.

A more general study of asteroid impacts on the earth's atmosphere has been published very recently by Jack Hills and Patrick Goda of the Los Alamos National Laboratory. They concur that the Tunguska evidence rules out a comet, and that furthermore there is no question but that the object was stony and not iron in nature. Irons can resist greater pressures to reach the earth's surface, but very large chondrites, intermediate in density between metals and ices, explode at altitudes like that of the Tunguska Event.

It was remarked many years ago that had the Tunguska object arrived at the earth only a few hours later, the earth's rotation would have brought the event over major European cities. As chilling as the thought is, even more chilling is the notion that yet a few more hours would have brought the event over the Atlantic. Hills and Goda, in the calm language demanded by the *Astronomical Journal*, write that

> an asteroid with a radius of [six hundred feet] that drops anywhere in the mid Atlantic will produce deep water waves that [traverse the open ocean at two hundred miles per hour and] are at least [sixteen feet] high when they reach both the European and North American coasts. When it encounters land, this wave steepens into a tsunami over [650 feet] in height that hits the coast with a pulse duration of at least 2 minutes, so it would move far inland before the pressure driving it stops. Low-lying areas such as Holland, Denmark, Long Island, or Manhattan may be washed out. (The numbers are very disturbing to the authors. Perhaps the legendary tale of the lost civilization of Atlantis, which was said to be on the Atlantic coast and to have been engulfed suddenly by the ocean, is attributable to such a tidal wave. It is somewhat surprising that there were no widespread coastal settlements along the Atlantic until after 800 A.D., when the Vikings settled and fortified numerous towns along the Atlantic coast. The niche that they exploited may have been opened by a previous disaster whose institutional memory had been lost.)

Perhaps we may draw some small comfort from an estimate by Sekanina: Events like the Tunguska collision probably occur only once in every two thousand to twelve thousand years.

13

Comets—Again!

I hope that somewhere along the line I get to see a comet that really turns me on. Things haven't gone well so far. True, when the comet of the century is announced every few years, I dash out to the backyard and stare through my binoculars like everyone else, but even when I actually succeed in finding the damn thing, I'm never really overcome by much more than ennui. Comet Halley—which, heaven knows, I've lectured about more times than I care to remember—drove me to use a professional telescope to pin it down when it came by in 1986, only to have my usually enthusiastic family dismiss that year's comet-of-the-century with a polite "Well, so that's it, eh?" My sentiments exactly!

Others have clearly thought otherwise. You may recall that comet Halley came by in 1066 when King Harold was battling the Normans at Hastings, an event—comet and all—commemorated on the famed Bayeux tapestry. Not the best of moments in English history. The comet itself as depicted on the tapestry looks rather like a child's toy with feathers, but in John Milton's later view it was something "That fires the length of Ophiuchus huge/ . . . and from his horrid hair/ Shakes pestilence and war," whereas Tennyson thundered, "You grimly-glaring, treble-brandished scourge of England."

Edmond Halley himself, of course, probably did rather well out of the comet. As noted in chapter 2, back in 1691, before he'd really got to work on comets, he applied for the Savilian Chair of Astronomy at Oxford, and was turned down in part because crusty old John Flamsteed, then astronomer royal, dismissed him as a "drunken sea-captain." It's true that Halley in his seafaring days had picked up at

First published in *American Scientist*, 82:104, January–February 1994.

least a modicum of quarterdeck language and habits while putting down mutinies and enduring the general vicissitudes of the seventeenth-century mariner's life, but by 1704, when he was really getting into the comet business, he found himself appointed Savilian Professor of Geometry.

History, though, seems to have generally taken comets rather somberly. As Ambroise Pare recorded in the early 1500s, "This comet was so horrible and so frightful and it produced such great terror in the vulgar that some died of fear and others fell sick. It appeared to be of excessive length and was the color of blood. At the summit of it was seen the figure of a bent arm, holding in its hand a great sword, as if about to strike. On both sides of the rays of this comet were seen a great number of axes, knives, and blood-colored swords, among which were a great number of hideous human faces with beards and bristling hair." I gather Pare didn't have as much trouble as I do in finding these things from the backyard.

Anyway, I see we're off again after the latest comet of the century. It involves nothing less than "what may well become the largest campaign in the history of modern astronomy." Moreover, "the event could have been the most spectacular astronomical event ever to be witnessed in the heavens during recorded history." (Comet people talk like that.) Sad about the past tense there, but we'll discuss that in a moment. Meanwhile, I'm relieved that I probably need not make that frustrating pilgrimage to the backyard this time.

The object in question is comet Shoemaker-Levy 9, named after the veteran comet hunters who discovered it in March 1993. (Carolyn Shoemaker has discovered at least thirty comets and is approaching the all-time record number of comets discovered by one person. David Levy offers outstanding testimony of the contributions of amateur astronomers to their subject.) Even on the small scale of the survey telescope, the comet appeared "squashed," and with a larger instrument it was resolved into some twenty components spread out in what was soon described as a "string of pearls." Significantly, Jupiter blazed brightly not far off in the sky.

Once the night-to-night motion had been measured, orbital calculations revealed that the comet had passed only some thirty thousand miles above Jupiter's surface on 8 July 1992. There it had been ripped to shreds by the enormous tidal forces of the giant planet and had become the latter's captive. The remnants now journey on a highly elliptical orbit about Jupiter itself.

This is not unique. More than a hundred years ago, comet Brooks 2 made a very close approach to Jupiter and was similarly shattered, although only a few faint companions of the main body were subsequently detected. What was unique about comet Shoemaker-Levy 9, however, became clear as further observations led to refined orbital calculations. The "pearls," traveling their extremely elliptical orbit, must eventually return to Jupiter again, and now the calculations revealed that they would come within twenty thousand miles of Jupiter's center of mass. Since the planet's radius is roughly forty-four thousand miles, collision with Jupiter is inevitable. This will be true for all the twenty or so components. On 17 July 1994, the string of pearls will spread across a quarter-degree of the sky (but will be invisible to us, being lost in the blaze of Jupiter) as, billions of years after its formation, the comet's remnants finally hurtle toward destruction. Sometime around late afternoon Greenwich time on 18 July, nucleus number 17 will head the lemminglike parade to fiery death in Jupiter's soupy atmosphere. One by one the nuclei will smash in, ending with nucleus 1 in the early morning of 23 July. All told, it will be an event that may happen only once in thirty thousand years.

Ah, but what about the past tense I noted earlier? There's the rub. The tragedy is that all this is going to happen on the far side of Jupiter, hidden from us by the giant itself. Moreover, just what does happen is critically dependent on the mass of the cometary fragments. Initially it was thought that the larger fragments might be as wide as six miles, in which case the effects of collision would be spectacular. Accelerated by Jupiter's huge gravity, the fragments are going to slam into the planet's atmosphere at around forty miles per second, and the largest might penetrate to atmospheric-pressure levels of around one terrestrial atmosphere. There they would experience aerodynamic pressures a thousand times higher, or some five orders of magnitude beyond their estimated material strength. The resultant explosion would release energy probably exceeding the equivalent of a million megatons of TNT. Some estimates run one or two orders of magnitude higher, rivaling the energy release of the Yucatan explosion at the end of the Cretaceous period. (But remember, Jupiter would experience a series of these, separated only by hours.) "Within a minute after [each] explosion a fireball as hot as the Sun will rise above the [Jovian] cloudtops." (More comet-people talk.)

Were the explosions on the side of Jupiter facing us at that moment, they would be visible from earth in broad daylight. But even though the impacts will be on the far side of Jupiter, a suitably placed Jovian satellite might reflect the brilliant flashes. All eyes will be watching for them. Within hours the rapid rotation of Jupiter will bring the impact zones into view, and there is the possibility that the impacts will have excited waves that cause Jupiter to "ring like a bell." If this produces cloud condensations visible to the Hubble Space Telescope, it may be possible to probe the structure of the atmosphere or even of the planet itself.

On the other hand (those dread words), such are the variables of cometary physics that if those fragments are only one mile rather than six miles across, the visible effects of the impact might be negligible. And indeed, the Hubble Telescope observations to date put the sizes at something like one or two miles. Some estimates put the parent itself at only two miles. (Photographs make the fragments look much larger, but that is because light is scattered by the dust surrounding each component.)

Such cometary excitement puts one in mind of E. E. Barnard's fabulous automated comet-hunting machine. Barnard, then a staff member at the Lick Observatory of the University of California, didn't know about it himself until he read the details in the *San Francisco Examiner* of 8 March 1891:

DISCOVERS COMETS ALL BY ITSELF
A wonderful Scientific Invention that will do away with the Astronomer's Weary Hours of Sweeping—It's Just Like Gunning for Wandering Stars with a Telescope.

Flabbergasted, Barnard read on to where he was supposedly quoted verbatim:

"Mark now the effect!" cried Barnard, almost rapturously: "When the moon goes down I will start the telescope 'sweeping.' I can then leave my comet-seeker to its own intelligent work, and give my attention to stellar photography and other important matters. Throughout the night my human telescope explores the skies; stars, nebulae, and clusters innumerable crowd into the field with every advance of the clock, but the telescope gives no sign of their presence, for the analyzing prism spreads out the light of even the brightest among them throughout the length and breadth of the spectrum, and when this spectrum falls on the three slits of the diaphragm its light is far too feeble to exercise any electrical effect upon the selenium.

"But let even the faintest comet come into range and see what are the consequences! The prism instantly analyzes the light, the bright hydrocarbon bands fall upon the respective slits. The light of these, reaching the strip of selenium, so changes the electrical resistance as to disturb the balance of the Wheatstone bridge, and a feeble current is sent through the wire. This in turn closes all the circuits of the powerful Leclanché battery, and the comet is caught, as in a trap.

"An alarm-bell rings in my bedroom down at the cottage."

Barnard, of course, leapt for paper and pen and wrote a white-hot denial to the *Examiner*. That was when the second surprise came. The hoaxer who had sent in the phony article had forewarned the paper that Barnard, in his modesty, would deny the story, and that the editor should ignore all protestations of innocence. Poor Barnard. In the absence of published denials, he was deluged for years by inquiries from people wanting to build themselves such machines.

Eventually, after the *Examiner* published an apology on 5 February 1893, the story was revealed to have originated with a junior assistant at the Lick Observatory, Charles Hill.

Barnard, however, had already done well out of comets. He himself related how in 1880 a Mr. Warner had offered two hundred dollars for every unexpected comet discovered by an American or Canadian. The first time Barnard won this prize, he and his wife decided to use it as a down payment on a house. Barnard turned to comet hunting with new enthusiasm, until eventually, as he said, "It came about that the house was built entirely from comets. This fact goes to prove the great error of those scientific men who figure out that a comet is but a flimsy affair after all . . . for here was a house, albeit a small one, built entirely out of them. True, it took several good-sized comets to do it, but it was done nevertheless."

The Shoemaker-Levy team should be so lucky!

14

In Pursuit of Vulcan

At the age of thirty-five, Urbain Jean Joseph Le Verrier was on the crest of a wave. Starting as a minor employee of a tobacco company, he had turned to chemical research with Gay-Lussac, and had now achieved worldwide fame as the codiscoverer of the planet Neptune. Codiscoverer not in the sense of having put eye to telescope and seen Neptune, but as nothing less than one who had carried out an incredibly difficult year-long mathematical calculation to predict where in the sky the hitherto unseen planet would be found, and having had others find the planet close to that spot. (The "co-" in "codiscoverer" opens a fascinating story I shall have to address another time. [See chapter 15.])

Now, incumbent of a chair of celestial mechanics (and later a chair of astronomy) created specially for him at the Sorbonne in Paris, he looked about him for new worlds to conquer. The year was 1847. Le Verrier conceived the project "of embracing in a single work the whole of the planetary system." By this he meant a study of all the major planets in the solar system, systematically combining their masses, motions, and mutual perturbations into a uniform whole that would satisfy observation. Fresh from Neptunian triumphs, he wished to "put everything in harmony if possible and, if it is not, declare with certainty that there exist still unknown causes of perturbations, the sources of which would then and only then be recognizable." It was not a project for the weak of spirit. It would occupy the rest of Le Verrier's life, not to mention four thousand pages of the *Annales de l'Observatoire de Paris*. (Le Verrier was appointed director of the Paris Observatory in 1854.) For starters, just as a tool, he established the expansion of the perturbing force to the seventh power of the

First published in *American Scientist*, 82:412, September–October 1994.

eccentricities and inclinations, including 469 terms dependent on 154 special functions.

Thus armed—along, one supposes, with his log tables—Le Verrier started out on the solar system from inner planets to outer ones. And damn!—wouldn't you know it?—ran headlong into problems with the very first case: Mercury.

I expect he wasn't too surprised. Mercury had long been a trouble-maker. Almost from Newton's day, celestial mechanicians had found themselves unable to reconcile theory with observation where Mercury was concerned. In particular, the transits of Mercury had been problematic. With Mercury between the earth and the sun, it is sometimes seen projected as a small black blob against the solar disk, slowly moving from edge to edge as the various motions involved change the picture. The trouble was that the predicted times of beginning and ending of these transits generally showed departures from those observed. Le Verrier undertook a new analysis.

Le Verrier, says one biographer, conducted his projects like battles—in which, however, he displayed more emotion than strategy. Not yet over the thrill of Neptune, he found a ready solution to his problem. Neptune had been found because it caused previously unexplained discrepancies in the orbit of Uranus. Could not the discrepancies in Mercury's motions be similarly explained? Indeed they could! Mercury's problems, proclaimed Le Verrier to the Academy of Sciences in Paris on Monday, 12 September 1859, came down to an unaccounted-for thirty-eight seconds of arc per century in the advance of its orbit's perihelion, and this was most likely caused by the gravitational effects of an as yet undiscovered planet between Mercury and the sun. Simon Newcomb in 1882, with a better knowledge of planetary masses, managed to aggravate the problem by changing Le Verrier's thirty-eight arcseconds to a now-famous forty-three arcseconds per century.

A word of explanation. Planetary orbits are ellipses with the sun at one focus—that is, on the major axis of the ellipse, but off-center. There is thus a point on the orbit, known as perihelion, where planet and sun are closest. Because of the gravitational perturbations from other planets, the orbit, and thus perihelion point, slowly slews around, that is, advances. In the case of Mercury, the observed advance is a huge 5,600 arcseconds per century, but 5,026 of this is due simply to

the earth's motion (precession of the equinox). Another 531 are predicted by Newtonian theory using modern values of the planetary
masses, leaving Newcomb's forty-three arcseconds per century unaccounted for. But put another planet into the system, one with just
the right combination of mass and orbit, and those extra 43 arcseconds
would be merely an addition to the Newtonian collection. The problem would be gone.

Had Le Verrier done it again? The world, or at least the astronomical world, waited with bated breath. Surely the question would
not take long to answer, because the necessary planet would have to
be even nearer the sun than Mercury, where it would revolve about
the sun so quickly that its transits of the solar disk would be relatively
frequent. Was there no record of such a transit?

Indeed there was! On 22 December, scarcely three months after
his announcement to the academy, Le Verrier received a letter from
a Monsieur Lescarbault, a physician in the rural district of Orgères,
announcing that on 26 March of that year he had seen just such a
transit of a dark body across the sun. Le Verrier was outraged. How
dare so important an observation have gone so long unannounced?
With Christmas hardly past, he rushed to the country to confront
Lescarbault.

Richard Proctor, doyen of Victorian astronomical gossips, relates
the ensuing interview with relish—and, doubtless, embellishment.
He quotes a witness:

> One should have seen M. Lescarbault, so small, so simple, so modest,
> and so timid, in order to understand the emotion with which he was
> seized, when Le Verrier, from his great height, and with that blunt in-
> tonation which he can command, thus addressed him: "It is then you,
> sir, who pretend to have observed the intra-mercurial planet, and who
> have committed the grave offence of keeping your observation secret
> for nine months. I warn you I have come here with the intention of
> doing justice to your pretensions, and of demonstrating either that you
> have been dishonest or deceived. Tell me, then, unequivocally, what
> you have seen." The lamb, trembling, stammered out an account of
> what he had seen. He explained how he had timed the passage of the
> body across the sun. "Where is your chronometer?" demanded Le
> Verrier. "It is this watch, the faithful companion of my professional jour-
> neys." "What! With that old watch, showing only minutes, you dare talk

of estimating seconds. My suspicions are already too well confirmed."
"Pardon me, I have a pendulum that beats seconds." "Show it me." The
doctor brings down a silk thread to which an ivory ball is attached, he . . .
shows his pendulum beats seconds.

Hardly mollified, Le Verrier inspected the doctor's telescope and
noted disparagingly that the observation had been recorded on a scrap
of paper covered with grease and laudanum, currently serving as a
bookmark. Had the miserable doctor attempted any calculations as
to the planet's orbit? He had, and, being both physician and carpen-
ter, he rummaged around in his workshop to find the boards on which
he had recorded the calculations with chalk. Le Verrier dismissed
them contemptuously.

A fine display of arrogance, but evidently Le Verrier was secretly
very pleased with the report. One can hardly imagine the surprise of
the trembling lamb when, after due interrogation of the town's mayor
as to his victim's general character, Le Verrier arranged for Lescarbault
to receive the Legion of Honor. Ah, mon Dieu, these astronomers. . . .

And so began Le Verrier's crusade to find the putative planet Vulcan
(a name proposed by Le Verrier's friend, Abbé Moigno), a crusade
he would pursue unfulfilled to the grave. At first, however, things
went well. In Le Verrier's expert hands, Lescarbault's observation
yielded seemingly reasonable, if somewhat uncertain, parameters for
Vulcan's orbit. Applause was widespread. "The singular merit of
M. Lescarbault's observations," wrote the *Spectator*, "will be recognized by all who examine the attendant circumstances; and astronomers of all countries will unite in applauding this second triumphant
conclusion to the theoretical enquiries of M. Le Verrier."

Unfortunately, the merit of M. Lescarbault's observations was rather
less singular than expected. An astronomer by the name of Liais, attached to the Brazilian Coast Survey, wrote to say that he had been
observing the sun at precisely the same time as Lescarbault and with
a better instrument, yet had seen nothing resembling a planet in transit. His honor impugned, the trembling lamb became more lionlike,
and a brisk exchange of views as to one another's abilities and characters broke out. Factions began to form. Le Verrier stuck by Lescarbault; Flammarion and others supported Liais in scorning Vulcan's
existence. Most waited, bemused, for the next predicted transit, which

was due sometime between 25 March and 10 April of the coming spring. Airy, England's astronomer royal, mindful that the sun never set on the British Empire, arranged for observations to be made almost continuously from the Orient, through India and Africa and Europe, to the Americas. Observers everywhere stood by their telescopes, ready to skewer Vulcan. The results? All negative. Nothing was seen.

Many an astronomer was not surprised. From Le Verrier's first announcement of Vulcan, there had been difficulties. Lescarbault's observation had implied quite a small object, and if it had a density similar to that of the other inner planets, its mass would have been quite incapable of affecting Mercury's orbit to the extent necessary to account for the missing forty-three arcseconds. On the other hand, if it were large enough to do the job, it was almost incomprehensible that it had not long since been found. Perhaps more than one small object was involved.

Le Verrier did not give up. His clarion call having gone out, he was for the rest of his life deluged with reports from around the world announcing observations of Vulcan. Those which seemed more reliable he would subject to careful analysis, but none ever gave success in predicting another transit. Most were probably honest enough, but were made by inexperienced people who were confused by sunspots.

Urbain Le Verrier, ultimately a disappointed man, was only sixty-six years old when he succumbed to a painful and protracted liver disease; he died on 23 September 1877, his nearly twenty-year search unfulfilled. Ten months later the world would be electrified by the news that two Americans had independently discovered Vulcan.

There was more than one way to find Vulcan. Transits across the sun's face, predicted to be several times a year, were the obvious way. But the occasion of a total solar eclipse was another. Normally one might have great difficulty seeing Vulcan if it were always so close to the sun in the sky, but at a total solar eclipse, with the moon hiding the brilliant disk, who knew what might appear alongside it? The eclipse of 29 July 1878 seemed tailor-made for the test. The path of totality swept down out of Canada and along the spine of the Rocky Mountains, affording untold miles of high, clear-air sites for observers. Preparations were extensive; famous scientists came from afar. Names included Charles Young, Sir Norman Lockyer, Samuel Langley, James

Watson, Simon Newcomb, and even Thomas Edison. Spectacular fourteen-thousand-foot peaks surrounded them, awaiting the rush of the moon's shadow.

At Separation, Wyoming Territory, Watson was among the astronomers poised for action. As the final seconds prior to totality ticked away, the harsh command of "Silence!" rang out, and the observers bent to their instruments. Darkness came, and, eye to micrometer eyepiece, the measurements began. With only two or three minutes for the task, many an astronomer never looked up to see the stunning sight. Watson worked furiously, measuring the angle between the sun's limb and various bright objects nearby. "I found, about a minute before the end of totality, a star . . . which immediately arrested my attention from its general appearance and place, in which there is no known star. It had a disk larger than the spurious disk of a star." Watson began to run, shouting to Newcomb as he went. But the going was rough, and Newcomb still unhearing, when the first blinding light of the sun's photosphere flashed across the scene.

The lunar shadow raced on at over a thousand miles an hour, reaching Denver some four minutes later. There Lewis Swift, knowing nothing of Watson's observation, would claim a similar sighting "of a celestial object not down in Argelander's charts, which to my mind, without any doubt, is the long-sought Vulcan."

Word was telegraphed around the world; the press was ecstatic. Mouchez, successor to Le Verrier in Paris, acclaimed the result and said it consecrated the glorious scientific endeavors of Le Verrier. Watson respectfully submitted a donation for a statue in honor of Le Verrier.

But the spoilsports were not to be denied. Watson and Swift did not agree on the position of the supposed object, despite corrections to the raw observations. Why had the many other experienced observers not seen Vulcan? Soon astronomers, particularly European astronomers, were dismissing the entire claim. They felt sure Watson and Swift had become confused in the rush of the moment and had seen only stars. The anguished cries of the two reverberated through the journals, and exchanges became increasingly bitter, although no one seems to have seriously believed that either Watson or Swift were guilty of any deliberate falsehood; they were simply victims of their own unconscious desires.

The skeptics prevailed, however. Once photography superseded micrometer measurements at solar eclipses, the conclusion was undeniable. By the turn of the century, virtually no one among the cognoscenti believed in Vulcan.

With Vulcan gone, though, what was to be made of the wretched forty-three arcseconds per century? Simon Newcomb, in his 1882 review, considered various possiblilites. Was the sun nonspherical, thus altering its attraction on Mercury? No, or at least not enough to account for the forty-three arcseconds without upsetting observations of the transits of Venus. Was there an intra-Mercurial asteroid belt? No, its necessary inclination would make it unstable. A disk of dust around the sun (Henri Poincaré's suggestion)? No, the zodiacal light refuted it. Was Venus more massive than thought? No, it would lead to excessive perturbations of the earth. Newcomb gloomily concluded that it might be necessary to fall back on Asaph Hall's suggestion that Newton's law of gravity was force diminishing not as the inverse square of distance, but instead as distance to the inverse power of 2.0000001574. Somehow it didn't seem like God's style.

More than thirty years later, Albert Einstein arrived on this scene. On 15 December 1915, he wrote to his friend Wladislaw Natanson on the subject of the new theory of general relativity: "I am sending you some of my papers. You will see that once more I have toppled my house of cards and built another; at least the middle structure is new. The explanation of the shift in Mercury's perihelion, which is empirically confirmed beyond a doubt, causes me great joy. "

The agreement between general relativity and observation is indeed good. A modern analysis by Narlikar and Rana finds that the observed excess in the shift over Newtonian theory is 42.99 arcseconds; relativity theory predicts 42.98! No doubt Le Verrier himself would be satisfied.

The Neptune Affair

Undoubtedly one of the most interesting questions about science is why scientists sometimes behave in stunningly unexpected and inexplicable ways. How is it that people trained to be dispassionate and disinterested (though never uninterested) in what they do, can all too often reveal themselves as wildly biased, even when the outcome means little to them personally? There can be few cases more curious in this regard than the discovery of the planet Neptune, one of the grand yet notorious astronomical events of the mid-nineteenth century.

The events that became so contentious happened in the mid-1840s, but the story began long before that. On 13 March 1781, William Herschel, whom we met in chapter 10, accidentally came across what he thought to be a comet, but which soon proved to be the first planet ever actually discovered. It was later named Uranus. As soon as a preliminary orbit had been established for Uranus, it became possible to extrapolate backward in time to see where the planet had been in recent decades and thus to check whether it had been mistakenly catalogued as a star by other observers. (Herschel had excellent telescopes, which revealed the planet's tiny disk; others, using lesser instruments, might easily have thought it a star.) The search proved fruitful, and soon a considerable number of such instances were available to improve the orbital calculation. With these, and the observations that became routine after 1781, one would have expected an increasingly precise determination of Uranus's orbit. Instead, a major problem was slowly revealed.

Alexis Bouvard, writing in 1821, explained that if one determined

First published in *American Scientist*, 83:116, March–April 1995.

an orbit from the earlier observations, it proved incompatible with later observations, and vice versa. No single set of orbital elements would fit *all* the observations, even with due allowance for the effects of Saturn, Jupiter, and the inner planets. Moreover, the errors were large and definitely significant. With much misgiving, he decided to assume that the early data were inferior and determined an orbit from recent data only. "I leave to the future," he wrote, "the task of discovering whether the difficulty of reconciling the two systems is connected with the ancient observations, or whether it depends on some foreign and unperceived cause which may have been acting upon the planet."

The future made up its mind pretty quickly, for it was soon found that Bouvard's own calculations failed to satisfy ongoing observations. The question was what foreign and unperceived cause was at work here. There were a few wild speculations about Uranus being battered by hordes of comets, or the inverse-square law of gravity possibly being inexact at Uranus's great distance from the sun, but the clear favorite among foreign-cause speculators was that there existed yet another planet beyond Uranus perturbing the latter gravitationally. Easy to say, but quite another thing to prove.

It is a straightforward procedure to calculate the perturbing effect of one body on another, given the orbits, positions, and masses of the pair, but the inverse problem of finding the mass and orbit of an unseen perturber is, strictly speaking, intractable. For one thing, the perturber might be of lower mass and nearby, or it might be of higher mass and farther away, or of any appropriate combination of mass and distance. Fortunately, though, there exists among the planets a curious relation known as the Titius-Bode law. It is not a law in the usual sense, but a simple empirical recipe for remembering the distances of the planets from the sun: in the series 0, 3, 6, 12, 24, 48, 96, 192, . . . , add 4 to each term and divide by 10. The resultant series, 0.4, 0.7, 1.0, 1.6, 2.8, 5.2, 10.0, 19.6, may be compared to the actual distances of the planets from the sun expressed in terms of the earth's distance as unity: 0.39, 0.72, 1.00, 1.52, 2.9, 5.20, 9.55, 19.2. The value at 2.8 between Mars and Jupiter had already led to a successful prediction of what turned out to be the asteroid belt. Uranus is the planet at 19.2, so clearly a good first bet for the unseen perturbing planet would be $((2 \times 192) + 4)/10 = 38.8$ units from the sun. Moreover,

Kepler's third law, relating the orbital periods of the planets to their distances from the sun, then offered some hold on the motion of the mystery planet.

Even so, there were still many unknowns in the problem, and any attempt at predicting where in the sky the newcomer might be found would involve long and laborious calculations, with doubtless many false starts. In a computationally challenged age, the experts waved the problem away; there were more tractable projects that were more likely to succeed. Fortunately, undergraduate students are less experienced.

In 1841, John Couch Adams was an undergraduate at Cambridge University and by far the most talented mathematician of his class. (A contemporary later recalled meeting Adams when they both first entered the university to study mathematics. A short conversation about the subject left the contemporary utterly aghast; if this was the average new student, then perhaps he himself should leave at once! As it was, Adams eventually graduated senior wrangler, further ahead of the second wrangler than the latter was ahead of the bottom of the class.)

On 3 July of that year, Adams penned a note to himself: "Formed a design in the beginning of this week, of investigating, as soon as possible after taking my degree, the irregularities in the motion of Uranus, which are yet unaccounted for; in order to find whether they may be attributed to the action of an undiscovered planet beyond it; and if possible thence to determine the elements of its orbit . . . which would probably lead to its discovery."

With his B.A. behind him, the twenty-four-year-old Adams in 1843 set out on his odyssey. He made further simplifying assumptions: a circular orbit coplanar with that of Uranus. The results showed promise; even the simplified calculations diminished the errors in Uranus's orbit. To advance, however, more data were needed. Adams laid his case before James Challis, Plumian Professor of Astronomy at Cambridge, who obligingly requested the astronomer royal, George Airy, to make available the recent errors in Uranus's position. Airy immediately sent not only these but all the errors back to 1754. Adams returned to his task.

In September 1845, Adams presented Challis with a prediction of where in the sky the new planet might be found as of 1 October 1845.

We now know he was less than two degrees off. And here the unexpected aspects of the story begin.

The technique needed to search for the planet was straightforward. It would require the mapping of all stars in the region at one epoch, a wait of a week or two, and then a repeat of the mapping. Comparison of the two maps would reveal any change in position of an object, indicating that it was not a star but a planet with orbital motion. Indeed, if one were lucky and the planet of some size, one might detect it the first time round by its disk. But in this prephotographic age the task would be tedious; setting a telescope star-by-star, reading off the coordinates, writing it all down. Challis was unenthusiastic, even though Cambridge had just the telescope for the job. Instead, he sent Adams off to Airy with a letter of introduction, probably hoping that Airy might be interested enough to have the search made at Greenwich. But Airy was away when Adams arrived. Still, on his return Airy immediately wrote to Challis to say that "I am very much interested with the subject of [Adams's] investigations." Adams returned to Greenwich, found Airy absent again, and returned at 3:30 the same afternoon, to be met at the door by the butler, who announced that the astronomer royal was at dinner and could not possibly be disturbed. Adams left a manuscript summarizing his work, and returned to Cambridge.

Airy duly read Adams's manuscript—with interest, it seems—sending him a note to ask a question about one technical point. And it was here that everything went off the rails. Adams was surprised at the question, regarding it as something someone of Airy's talents should have seen the answer to at once. Thinking it was perhaps rhetorical, he let it pass without answer. Perhaps he didn't know that Airy was one of the most neurotic characters in all of astronomy. The result was that Airy wrote to Challis, "Adams's silence . . . was so far unfortunate that it interposed an effectual barrier to all further communications. It was clearly impossible for me to write to him again."

So it was that in late 1845 Adams, Challis, and Airy let the matter rest. The modern mind is staggered by the situation. Here was one of astronomy's greatest triumphs within their grasp, yet Airy was too proud to act, Challis too uninterested, and Adams himself too retiring to take even the simplest of actions. Scorn and anger would later be heaped on Airy and Challis, but what of Adams himself? If the first two had too little faith to think it worth the effort of scanning a small region of the sky to check the prediction, why didn't Adams, who had put in years of laborious calculation to arrive at his prediction, not do more? There he was, daily passing the ideal telescope for the job, yet nowhere in the considerable literature on this subject have I found any indication that he ever asked, let alone pleaded, to be allowed to do the simple search himself. Even allowing for stiff nineteenth-century protocol, it seems incredible that he merely sat back with a sigh.

A few months earlier the dazzlingly talented young Urbain Le Verrier (whom we met in the previous chapter on Vulcan) had entered the lists. Protégé of Dominique Arago, the president of the French Academy of Sciences, Le Verrier set out on just the same path Adams had already completed. But whereas Adams's work was known only to himself, Challis, and Airy, Le Verrier's work was published piece by piece in the *Comptes rendus* of the academy. With Neptune now in the daylight sky, the British trio must have followed with interest Le Verrier's homing in on the conclusion long since reached by Adams.

Airy put to Le Verrier the same question he had put to Adams. Airy was pleased to receive a satisfactory reply. He never mentioned Adams. But by the summer of 1846, Airy did see fit to urge Challis to start a search for the putative planet. Adams produced an update to the expected coordinates, and on 29 July Challis began to sweep the sky, writing down the coordinates of all the objects in the general area. By 13 August, he had accumulated four nights of observations, and he began to compare the 30 July measures with those of 12 August. He reached star number 39 on his list before ennui overcame him. Had he got to number 49, he would have discovered its motion, for it was Neptune. Indeed, had he earlier compared his observations of 4 August with those of 29 July, he would have discovered Neptune; its motion was already apparent. But the opportunity was lost.

Finally reacting to the situation, Adams in early September raced to a meeting in Southampton of the British Association for the Advancement of Science to read a paper on his work. He arrived to learn that the physical sciences section had concluded the previous day. It was now the turn of the biological sciences.

On 29 September, Le Verrier's final paper on the subject reached England. In it he gave the orbital elements of the new planet, its mass, and its coordinates (only a few degrees from Adams's prediction). He also noted that it would probably present a nonstellar disk to the observer, a conclusion already announced by Adams months earlier. Now finally galvanized, Challis got to work the same night, searching three hundred objects in the area for one with a sensible disk. Only one attracted his attention. It was indeed Neptune, but Challis decided to wait for confirmation on another night. The bright moon was in the area the following night, and the day after that he heard that Neptune had just been discovered in Germany.

Le Verrier had in fact run into much the same problem as had Adams. His French compatriots applauded his theoretical work, but no one raced to a telescope to verify it. Instead, he had to write to a friend, Johann Galle, at the Berlin Observatory to ask for observational help. Galle received the letter on 23 September and started the search the same night, assisted by a volunteer, Heinrich d'Arrest. They were fortunate in that the Berlin Observatory had recently produced a star map of the region, already printed but not yet distributed to other observatories. Thus they had to look only for an object not on the map. Hardly had they started when Galle at the telescope called out the coordinates of the current object, and d'Arrest, sitting with the chart before him, exclaimed, "It is not on the map!" The search was over.

Airy wrote to congratulate Le Verrier. He finally managed to announce that "collateral researches had been going on in England" without mentioning Adams by name or institution. "If in this I should give praise to others," he continued, "I beg that you will not consider it as at all interfering with my acknowledgements of your claims. You are to be recognized beyond doubt as the real predictor of the planet's place."

Astronomers all over Europe hailed Le Verrier as a hero who had accomplished one of the greatest triumphs in the history of mathematical astronomy. The French were ecstatic. Their celebrations, however, changed abruptly when a public announcement of Adams's work was made on 4 October by Sir John Herschel. As a member of the Board of Visitors of the Royal Greenwich Observatory, Herschel had caught wind of what had been going on—or rather had not been going on! Arago and his academy were wild with anger and outrage. Who was Adams? No one had ever heard of him! Were the unspeakable British trying to steal France's glory? A session of the Academy of Sciences on 19 October was so turbulent with violent rhetoric that a leading Parisian journal noted, "What a meeting! Are we in the Academy of Sciences or in the Chamber of Deputies?" Brutally satirical caricatures of Adams appeared elsewhere in the French press amidst denunciations of the British nation and its "odious national jealousy."

The British astronomical establishment was hardly less angry, and finally the astronomer royal and Challis were called to account. A meeting of the Royal Astronomical Society on 13 November 1846 received "the three most remarkable communications . . . which the Society can ever expect to receive in one night." Airy led off with his account of what had happened, followed by Challis and his side of the story, and finally Adams was at last able to present his own work. He was loudly applauded. Airy gave a cool account that made clear he saw himself as in no way to blame. He offered not the slightest apology. Challis, by contrast, simply caved in and admitted indifference and lack of confidence in Adams. Even a close friend of Airy's (how many could there have been?) accused him of having "snubbed [Adams] from the first . . . and so prevented him from reaping the honours of a great discovery." Airy remained completely indifferent to the attacks (although noting, "I was abused most savagely"), but Challis was broken and finally left Cambridge.

The uproar eventually died down, and even the French acknowledged Adams's work. Sir John Herschel, old smoothy that he was, addressed the Royal Astronomical Society in 1848 with the words. "As genius and destiny have joined the names of Le Verrier and Adams, I shall by no means put them asunder; nor will they ever be pronounced apart so long as language shall celebrate the triumphs of science in her sublimest walks." Le Verrier became director of the Paris Observatory, and Adams, fittingly, finally found himself with the Cambridge position previously held by Challis. The two, Adams and Le Verrier, became and remained firm friends for life.

There are several ironies to the story. After Neptune's orbit had been determined from further observations, it was found that the orbital elements departed considerably from those predicted by either Adams or Le Verrier, mainly because the Titius-Bode law does not hold well for Neptune. This led Benjamin Peirce of Harvard to claim that it was only by good luck that the planet had been found, a claim, of course, hotly rejected on the other side of the Atlantic.

Adams and Airy, not surprisingly, never warmed to each other, yet their backgrounds, abilities, interests, and careers were closely parallel, and nearly a half-century after the Neptune affair, they died only nineteen days apart.

The greatest irony, though, was one never known to any of the principals in this story. It was that an astronomer far more famous than any of them had had Neptune within his grasp without realizing it centuries earlier. In 1980, Charles Kowal and Stillman Drake realized that in December 1612–January 1613, Neptune had passed behind Jupiter just at a time when Galileo was studying the motions of Jupiter's satellites. A search of Galileo's observing log showed that he had drawn diagrams of the sky near Jupiter, with background stars included. The planetary ephemeris of the Jet Propulsion Lab, along with stellar catalogs, showed exactly what Galileo had seen through his telescope at the times and dates recorded in the log. Sure enough, on two nights Galileo included Neptune as a star in his diagrams (his telescope would not have resolved the disk), and he must have seen it on other nights. Presumably he was too absorbed in the Jovian satellites to notice that the background pattern was changing. Had he done so, Neptune would have been discovered nearly 234 years earlier.

16

The Great Debate

Astronomy has been more fortunate than most sciences, I suppose, in having from time to time been able to make society sit up and take note of some finding that would change forever its fundamental understanding of the world about us. It has rarely been easy. Copernicus springs to mind most readily, but even he made little impression in his day. As he lay on his deathbed clutching, perhaps indifferently, the recently published volume that would prove to be his masterwork, he was spared the knowledge that it would be largely neglected for decades to come. On the other hand, there was Kepler, rejoicing wildly over what he felt were discoveries that would set the world afire, never aware that history would consign them to the scrap heap while revering him for discoveries he himself had thought little of. No, rarely have the protagonists been able to foresee their place in history with any accuracy.

In our own century, with all due acknowledgment of the absolutely stunning discoveries that have come with the space age in which we live, I believe the greatest single leap in astronomical understanding came in the 1920s. Perhaps the only challenge would come in the discovery of the cosmic background microwave radiation and its fine structure, signatures of the universe's birth. But to the humanist in the street, it matters more whether our Milky Way galaxy exists alone in a boundless universe or whether the chance of life elsewhere is multiplied untold times by our galaxy being but one among immeasurable others. This year, 1995, marks the seventy-fifth anniversary—the diamond jubilee, it's been called—of what is known as astronomy's great debate, a confrontation of opposing experts on whether or not the Milky Way exists alone. The Shapley-Curtis debate took place

First published in *American Scientist*, 83:410, September–October 1995.

before the National Academy of Sciences in Washington, D.C., on 20 April 1920.

To commemorate it, on 22 April 1995, another great debate was held in the same Baird Auditorium where Shapley and Curtis battled it out before. Beginning with a retrospective of the original debate by Virginia Trimble, the 1995 program went on to a debate between Bohdan Paczyński of Princeton University and Donald Lamb of the University of Chicago regarding the nature of what astronomers call gamma-ray bursters—one of the great mysteries of current astronomy. Where before Shapley and Curtis argued over the distances of spiral nebulae, in particular whether, compared to the Milky Way, they are "coequal empires, or dependent but nearly self-governing colonies" (Eddington's phrase), Paczyński and Lamb argued over the distances and nature of objects that produce violent outbursts of gamma rays on a time scale of seconds but are undetectable at any other electromagnetic frequency. The possible range of distances is mind-boggling: anything from the fringes of the solar system to the distant reaches of the universe. Put another way, anything from a few light-hours to billions of light-years! I shall return to the Pacyński-Lamb debate in chapter 18; for now we'll take up the story that underlay the Shapley-Curtis confrontation.

In fact, it almost didn't happen. In 1919, George Ellery Hale, founder and director of what would later be the Mount Wilson and Palomar Observatories, had proposed that the National Academy should the following year hold a debate on some topical subject, possibly island universes (i.e., galaxies) or the hot new theory of general relativity. Hale himself was for general relativity, but was overruled by the academy's home secretary, Charles Greeley Abbot (of solar physics fame). The latter gloomily replied to Hale, "From the way the English are rushing relativity in *Nature* and elsewhere it looks as if the subject would be done to death long before the meeting of the Academy," going on to betray his own feelings: "I pray to God that the progress of science will send relativity to some region of space beyond the fourth dimension, from whence it may never return to plague us." So to Abbot we owe astronomy's Great Debate on the "island universe theory."

After some uncertainty, the debaters selected were thirty-four-year-old Harlow Shapley and forty-seven-year-old Heber Curtis. Shapley was then the bright young man (some would have said *enfant terrible*) of American astronomy. From a rural background in Missouri, where, after receiving a fifth-grade education, he had been a sixteen-year-old reporter for a small-town newspaper, he had by his early twenties gained entry to the University of Missouri, turned accidentally to astronomy, and won a graduate scholarship to Princeton. There, with Henry Norris Russell as supervisor, he completed in two years his doctoral thesis, 176 printed pages replete with ten thousand observations, a work that Otto Struve later called "the most significant single contribution toward our understanding of . . . very close double stars." It is a measure of the man, though, that this work is one of the lesser things for which he is remembered today.

After Princeton, Shapley quickly landed a job at the Mount Wilson Observatory in California, where Hale was director. It was here that the next few years would see Shapley catapulted to world fame. Indeed, although only thirty-two, he turned the astronomical world on its head with his analysis of the structure of the Milky Way.

A brief digression. In 1904, Henrietta Leavitt at Harvard had discovered that a class of variable stars called Cepheids have the property that their periods of variability correlate closely with their intrinsic luminosities. She had established this from observations of a group of Cepheids that are all at nearly the same distance from us, so the period-luminosity relation had shown up in terms of their observed brightnesses. Had she known the distance of the group, she could have calibrated the relation in terms of the intrinsic luminosities, and that would have provided a valuable tool. Thereafter, a measurement of a Cepheid's period would have yielded its intrinsic luminosity, which with the apparent luminosity would, through the inverse-square law, have told its distance—an immensely useful tool for mapping the universe around us. But Leavitt had not found a way to do this.

Shapley, once ensconced at Mount Wilson, had begun a massive attack on variable stars in star clusters, finding many of them to have light variations closely resembling those studied by Leavitt. What was exciting, though, was that this type of cluster, the so-called globular type, did not hug the plane of the Milky Way; instead, the clusters seemed to form a great halo around it. If he could find the distances of these clusters, Shapley realized, he could map out the extent of the Milky Way and our position within it. The Cepheid members of a cluster would be the key, if only he could find some way of calibrating Leavitt's period-luminosity relation. And indeed he did! Through an ingenious statistical procedure applied to a random group of nearby Cepheids, Shapley did manage to calibrate the relation. Applied to the cluster Cepheids, it revealed a picture of the galactic system that must have astonished even Shapley. Writing to Arthur Eddington at Cambridge University on 8 January 1918, he noted, "I have had in mind from the first that results more important to the problem of the galactic system than to any other question might be contributed by the cluster studies. Now with startling suddenness and definiteness, they seem to have elucidated the whole sidereal structure."

Prior to this moment, the best studies of the Milky Way system had suggested that it is a lenticular system of stars, some eight thousand light-years in diameter, with the solar system embarrassingly near its center. (What would Copernicus have said!?) Shapley's work, by contrast, indicated a Milky Way some three hundred thousand light-years across, with the sun a good sixty thousand light-years from the center! In his letter to Eddington, Shapley, with a sense of understatement that even the Englishman couldn't have bettered, noted that "this view of the general system, I am afraid, will necessitate alterations in our ideas of star distribution . . . in the galactic system." Years later, Walter Baade put it more boisterously: "I have always admired the way in which Shapley finished this whole problem in a very short time, ending up with a picture of the Galaxy that just about smashed up all the old school's ideas about galactic dimensions. It was a very exciting time, for these distances seemed to be fantastically large, and the 'old boys' did not take them sitting down."

It was in the wake of all this, then, that Shapley had been picked as one of the contenders in the Great Debate. One can only imagine Charles Abbot's thoughts on the choice.

By contrast, Heber Curtis, the other contender, was much more a background figure. A former professor of Latin and Greek at Napa College in California, he had become interested in astronomy and at the age of twenty-four had somehow made a smooth transition to a professorship of mathematics and astronomy. After acquiring a Ph.D. at the University of Virginia in 1902, he was invited to join the staff of the University of California's Lick Observatory, where he was later launched on a vast program of studying spiral nebulae. By the time of the Great Debate, many years of work and thought about these objects had convinced Curtis that they were galactic systems themselves, lying far beyond our own Milky Way. Shapley, on the other hand, having an enormously expanded view of the Milky Way's size and, as we now know, misled by a piece of totally erroneous evidence, believed the spirals were merely appendages of our system. This, indeed, was intended as the heart and thrust of the Great Debate. Was the Milky Way the only such object in the universe, or was it but one among a great many?

One would think that the talented, if seemingly brash, Shapley, already the center of a storm, might have relished the debate, while the older, quieter Curtis might have wished himself out of it. To the contrary, though, it was Shapley who had grave misgivings, while Curtis waxed enthusiastic. Shapley believed he was then the front-runner for the position of director of the Harvard College Observatory, and, with the debate taking place in nearby Washington, he feared a contingent of Harvard bigwigs would be there to see his performance against the experienced and accomplished public speaker that Curtis was known to be. What, thought Shapley, if he did badly? Curtis, unwittingly rubbing it in, wrote him, "I am sure that we could be just as good friends if we did go at each other 'hammer and tongs.'. . . A good friendly 'scrap' is an excellent thing once in a while; sort of clears up the atmosphere."

If there is one thing posterity has agreed on, it is that the versions of the two opposing arguments that were published a year later in the *Bulletin of the National Research Council* bore little resemblance to the spoken versions at the debate itself. For one thing, the nervous Shapley had (to Curtis's horror) managed to have the spoken arguments limited to thirty-five minutes apiece, while the later written versions would have taken over two hours to deliver. Moreover, Shapley's oral version was mainly at a popular level (presumably for the sake of any Harvard nonastronomical brass that happened to be there), while Curtis's efforts were technical and hard-hitting. Shapley emphasized that his vastly expanded Milky Way could accommodate considerable distances for the spiral nebulae, while Curtis argued that, expanded or not, it still could not contain the nebulae. So far as settling the issue went, the Great Debate was a bust. Pretty well everyone went home with their proclivities unchanged.

This is not the place to enter into all the technical arguments for or against the spiral nebulae being galactic systems themselves, but one—espoused by Shapley—was central. As noted in chapter 1, a colleague of Shapley's at Mount Wilson, Adriaan van Maanen, had been using their 100-inch telescope (the world's largest) to photograph some of the spiral nebulae. On comparing his plates with others taken decades previously, he concluded that these nebulae showed relatively rapid angular rotation. This could mean only that they were nearby, for otherwise their linear rotational speeds would be absurd. Shapley felt this must put the spirals within, or at least close to, the Milky Way. Curtis, on the other hand, while having no (publishable) argument against this finding, drew on other strong evidence that the spirals are far beyond our galactic system. The trouble was that no definitive evidence had been found. All rested on probabilities; loopholes abounded.

And so several years passed after the Great Debate before a new figure entered the scene. Four years even Shapley's junior, Edwin Powell Hubble arrived as the latest appointee to the Mount Wilson staff. His fascinating background has often been told, and is briefly summarized in chapter 1. At Mount Wilson, Hubble, like his colleague van Maanen, began studying photographs of spiral nebulae and found they contained Cepheid variables. With some sense of irony, no doubt, he used Shapley's calibration of the period-luminosity law to prove that the spirals are at distances far beyond the Milky Way, in fact, that they are galaxies themselves. The debate was finally over.

So how did it all turn out in the end? Shapley discovered after the debate that he had not been a contender for the Harvard directorship after all; that had been intended for his mentor, Russell. But Russell declined the offer and recommended Shapley in his place, so Shapley ended up as Harvard director anyway. Curtis, in the same year as the Great Debate, accepted the directorship of the Allegheny Observatory at the University of Pittsburgh, where he settled into administrative work and never again emerged into the realms of research in which he had once excelled. Of van Maanen it might be said that he became a casebook study in the sociology of science. Remeasures of his photographs by his colleagues revealed no measurable rotational motions in the spirals at all. Yet it can only have been self-delusion, an overriding wish to have the evidence conform to his views.

Hubble, of course, went on to ever more fame, although it is said he never got on with Shapley. They had their roots not many miles apart in rural Missouri, but while Shapley titled his autobiography *Through Rugged Ways to the Stars*, Hubble affected an Oxford accent, smoked a pipe, and wore leather pads on the elbows of his tweed jackets. Later though, when he had discovered the expansion of the universe (and, more importantly, got his photograph on the cover of *Time* magazine), yet another aspect of this remarkable man emerged. As noted in chapter 1, various Hollywood agents, not far from Hubble in Pasadena, thought fit to bring aspiring movie stars to be photographed alongside the great man. This on occasion backfired with male stars, for Hubble was not only a great scholar and athlete but also astonishingly handsome. A memorable group photo shows a puny Raymond Navarro, a current heartthrob, alongside the tall, good-looking Hubble. Women swooned; Navarro raged. In the end though, Hubble was only in his early sixties when, athlete though he was, he succumbed to a heart attack.

Perhaps the Great Debate was only a milestone down the road of history, inconclusive at that, but it served to concentrate and focus an important issue. Its significance has steadily grown with the passing years.

The Extraordinary and Short-Lived Career of Jeremiah Horrocks

At some point in my far-off adolescence, I became absorbed in Thomas Gray's poem "Elegy Written in a Country Churchyard." Aside from the natural beauty of its language ("The curfew tolls the knell of part-ing day"), I was taken with the lines

> Full many a gem of purest ray serene,
> The dark unfathom'd caves of ocean bear:
> Full many a flower is born to blush unseen,
> And waste its sweetness on the desert air.

Perhaps it was my growing up in South Africa, so far removed from Gray's eighteenth-century English churchyard, that touched a chord. About me was a vast sea of black faces, among whom must be all kinds of singular talent and ability, that had been born, would live, and eventually die, unrecognized because the darkness of their skin had rendered these abilities invisible to my race, then the white mas-ters. Happily that is now changing. But the words come back to me whenever the name of Jeremiah Horrocks is mentioned.

Jeremiah Horrocks was an Englishman whose talent possibly ri-valed that of Newton, and who died, aged twenty-two, the year be-fore Newton was born. We have no idea what he looked like and are uncertain as to his parentage and the date of his birth, even the spell-ing of his name, what he did for a living, or how he came to die, except that it was "suddenly." Most of what he wrote in his short lifetime is lost, but what remains bespeaks an astonishing mind and ability. Newton himself, when eventually writing his great work, the *Principia*, acknowledged his debt to "Our Countryman Horrox."

Horrocks was born in Toxteth, Liverpool, probably in 1619, al-

First published in *American Scientist*, 84:114, March–April 1996.

though no record of his birth survives. "His mother," says Allan Chapman, who likely knows as much as anyone about Horrocks, "was Mary Aspinwall, though as to whether his father was called William or James, there is no exact record." Jeremiah's early childhood is a blank; he emerges at about age thirteen entering Cambridge University as a sizar—a poor scholar who must work as a college servant to earn his keep. Perhaps the rising of Jeremiah to the status of a Cambridge student from an obscure family-of-the-land already implies very unusual abilities, although the age of thirteen was not particularly extraordinary for the times.

The Cambridge syllabus would have been Aristotelian, and certainly would have included nothing pertaining to the astronomy of the contemporaneous Copernican revolution. That would have been strictly a matter of self-education, although clearly it fascinated Horrocks.

It is, in fact, important to realize that Horrocks lived in a critical era of astronomical history. Johannes Kepler, something of a hero to Horrocks but not to many others, had died when Horrocks was eleven, while Galileo would outlive Horrocks by a year or so. Kepler, today a towering hero of elementary textbooks for his three famous laws of planetary motion, was still pretty much a nonentity. Well over a hundred years would pass before anyone (Voltaire, of all people) would refer to even one of them as a "law," and nearly two hundred years would pass before all three were gathered together in a textbook as "Kepler's three laws." In Horrocks's time, it was rather the view that anyone who claimed to displace the two-thousand-year-old scheme of perfect circles and constant speeds for the planets must be highly suspect, especially one who made such wild claims for musical harmonies and Platonic solid spacings among the planets. If nothing else points to Horrocks's insight, it is his realization that Kepler was on the right track. And it was largely Horrocks's enthusiasm for Kepler that later led Newton to ponder the seemingly arbitrary Keplerian results on planetary motions; surely some more basic law underlies them? Indeed it does—the Newtonian law of gravity.

Quite likely it was Horrocks's impoverishment that kept him from ever earning a degree, when, as Chapman says, "It was an expensive business going through the rituals of 'supplicating' for a degree . . . when special caps, gowns and hoods had to be purchased and University officers liberally tipped. Many men, upon completion of their studies, left without one, intending to return when they had made some money to be examined, and thereby regularize their status as Bachelors and Masters of Arts."

But so far as we know, Horrocks never returned to Cambridge once he had left. Instead we find him ensconced at about age twenty in the village of Hoole in Lancashire. Here, popular and persistent legend has it, he was a curate, a kind of junior, assistant priest in the Church of England. This, Chapman shows, is most unlikely. At the very least, Horrocks was too young for such a post; he lacked the usually necessary M.A., and moreover, none of his friends and contemporaries ever referred to him as having the title of "Reverend." Much more likely, says Chapman, he was a schoolmaster.

The popular assumption that he was a priest rests largely on an entry in his diary for 24 November 1639. This was a Sunday (in the Old Style calendar then in use in England), and he was anxious to observe a so-called transit of Venus but was called away "by business of the highest importance which, for these ornamental pursuits, I could not with propriety neglect." It's true, one would be hard-pressed to come up with any pursuits which a schoolmaster might not neglect on a Sunday (although he might also have been a Sunday school teacher), but the bulk of the evidence favors Chapman. In any case, it is the transit, not Horrocks's status, that interests us.

A transit of Venus refers to the fact that because Venus lies between us and the sun, and because the planetary system is nearly coplanar, there must be times when sun-Venus-earth will be a straight line, and observers on earth will see Venus projected against the disk of the sun, slowly transiting it as the motions of the two planets alter the observer's relative position. The crux lies in the word "nearly." The orbit of Venus is tilted to that of the earth by more than three degrees, and since the sun's disk covers only about half a degree on the sky, most often the line of sight to Venus will miss the sun's disk altogether. Kepler had decided that there would be only one transit every hundred years or more, the nearest being in 1631 (the year, it turned out, following Kepler's premature death), but Horrocks's reworking of the problem revealed that transits come in pairs eight years apart. Thus there would be transits in 1631 and 1639, then in 1761 and 1769, in 1874 and 1882, in 2004 and 2012, and so forth. By stunning good luck, Horrocks arrived at this result only weeks before the 1639 transit was due, and, unlike the 1631 transit, it was to take place during daylight hours in Europe, whereas the 1631 transit had happened while the sun was below the European horizon.

But why the excitement? At the time there was no reliable estimate as to the scale of the solar system, and Kepler had pointed out that if Venus is comparable to the earth in linear size, its angular size estimated against the half-a-degree disk of the sun would provide a much better estimate of that solar system scale. In fact, if two observers widely separated on earth were to observe a transit and find that the observed paths across the sun's disk differed slightly, it should be possible to determine the solar system's scale without making assumptions. Under the circumstances, of course, there was no hope of Horrocks arranging this, but he was very keen to have a go at estimating the angular size of Venus; as far as he knew, there wouldn't be another chance for the next 122 years!

An unhappy thought, though, was that the weather of an English November might well cloud the transit from view. To minimize this, Horrocks had enjoined a friend, William Crabtree, who lived some thirty miles away, to observe the transit as well. The pair, in fact, had never met (and never did; Horrocks died the day before their intended first meeting), but kept up a keen correspondence. Crabtree was a linen draper in Manchester, but one who, said Horrocks, "in mathematical knowledge is inferior to few."

Horrocks's prediction was that the transit would happen in the late afternoon of Sunday, 24 November, but so uncertain were the parameters in the calculation that Horrocks got in some observing the day before, and got an early start on the twenty-fourth just in case. He used a small telescope pointed at the sun, which projected the solar image on a screen in a darkened room. "I watched carefully and unceasingly for any dark body that might enter upon the disk of light," he tells us. The hours passed; the day became cloudier. No Venus. There were interruptions when—that famous sentence of his—he "was called to business of the highest moment." (What *could* it have been to outweigh Venus on that day?) "But in all these times," Horrocks continues, "I saw nothing on the Sun's face except one small and common spot, which I had seen on the preceding day." The day was waning and sunset just a half-hour away, when "the clouds, as if by"*Divine Interposition*, were entirely dispersed . . . and I then beheld a most agreeable sight, a spot, which had been the object of my most sanguine wishes, of an unusual size, and of a perfectly circular shape, just wholly entered upon the Sun's disk. . . . I was immediately sensible that this round spot was the planet *Venus*, and applied myself with the utmost care to prosecute my observations."

Crabtree, like his friend, saw the start of the transit, but with a less successful outcome:

> He eagerly betook himself to his observation, and happily saw the most agreeable of all sights, *Venus* just entered upon the *Sun*. He was so ravished with this most pleasing contemplation, that he stood for some time viewing it leisurely, as it were; and, from an excess of joy, could scarce prevail upon himself to trust his own senses. For we astronomers have a certain *womanish* disposition, distractedly delighted with light and trifling circumstances. . . . Which levity of disposition, let those deride that will. . . . For what youth, such as we are, would not fondly admire *Venus* in conjunction with the *Sun*, what youth would not dwell with rapture upon the fair and beautiful face of a lady, whose charms derive an additional grace from her fortune?—But to return, he from his ecstasy, I from my digression. The clouds deprived Mr *Crabtree* of the sight of the Sun, almost as soon as he was roused from his reverie; so that he was able to observe little more than that *Venus* was certainly in the Sun.

They were indeed youths, young men in their twenties.

And what of the results? Horrocks, carefully drawing the image of Venus against that of the sun on the screen, concluded that the planet had subtended an angle of no more than 0.02 degree. Crabtree, hastily making amends and instantly drawing Venus from memory, ironically arrived at a figure of 0.0175. I say ironically because I reproduced all the details of this transit on my computer, and the software tells me that Venus would have subtended an angle of just 0.0175 degree!

To understand the significance of this, one must realize that it implied a size for the solar system that was almost an order of magnitude greater than the popular figure of the time. Even so, this meant the sun was only sixty million miles away. It would not be until the next transits of Venus, more than 120 years later, that the figure closed in on the modern value, near ninety-three million miles. Nevertheless, by noting the precise time of the event, Horrocks was able to greatly improve the parameters of Venus's orbit.

Horrocks wrote this all up in a work that would twenty-three years later be published, not in England but on the Continent, under the title *Venus in sub sole visa* (Venus in transit before the sun).

If Horrocks is recognized at all today, it is usually regarding that transit of Venus. But we can see that his abilities far transcended that. For one thing, like Newton, he had a stunning ability to always ask the right question, and while he was a man of great theoretical insight, his dexterity of hand (and eye) was, again like Newton's, astonishing. Horrocks devised new instruments that might have changed the astronomy of his day had word of them got about. Yet here another irony intrudes: the greatest advance in instruments of that astronomical age was developed by a third character in the story, another friend of William Crabtree, by name William Gascoigne. It was he who first developed the means of measuring small angles by using movable cross hairs within a telescope, the so-called filar micrometer. Had Horrocks known of this earlier, he could have measured the angular diameter of Venus without need of a transit. As it was, he heard about it from Crabtree only in the concluding months before his death. Word of it, we hear, "ravished" Horrocks's mind, making Crabtree ask Gascoigne if he "could purchase it with travel, or procure it with gold." No evidence exists, however, that Horrocks ever laid hand on a filar micrometer in those remaining months.

On the theoretical side, one must reflect on Horrocks's study of Kepler. The latter has been hailed as the first astrophysicist, meaning that he was the first significant figure in history to ask what physical forces cause the planets to act as they do. All those before him—Aristotle, Ptolemy, even Copernicus—had been content with a purely kinematical scheme. If a planet went at such a speed in such an orbit, then it would appear as we see, but as to what forces would invoke this, no one before Kepler had ever considered. Aware that William Gilbert had recently shown the earth to be a vast dipole magnet, Kepler was all for a magnetic force emanating from the sun. In a scheme that is now seen to be naive, he proposed a stratagem whereby the presumably rotating sun would at times push a planet away, at times draw it closer, thereby producing an elliptical rather than a circular orbit.

Horrocks, however, in his study of the subject came to a different conclusion: while he couldn't yet say what the force was, it seemed it must always be an attractive force, as Newton would conclude years later. Horrocks linked this to the principles of falling bodies on earth, noting "Ye suns conversion doth turn the planet out of this line framing its motion into a circular, but the former desire of ye planet to move in a streight line hinders the full conquest of ye sun, and forces it into an Ellipticke figure." One can hardly come closer to Newton's laws of motion! And before that, he had made a careful study of the lunar orbit, concluding that the moon too obeys Kepler's laws in revolving about the earth. Thus, whatever that attractive force was, it was not unique to the sun.

All this from someone scarcely beyond adolescence. What might have emerged had Horrocks survived and come to know Newton, only twenty-two years his junior? Chapman, however, rightly cautions against such speculation. Who knows where Horrocks's interests might later have taken him; perhaps it was in a way convenient for the hero-worshippers of the eighteenth and nineteenth centuries that he died so young. Nevertheless, he stands acknowledged as the first important English figure in the new astronomy.

Nearly all the characters in this story died young: Horrocks suddenly the day before his first meeting with Crabtree, for reasons unknown, Crabtree two years later, again for reasons unknown, and Gascoigne in the same year during a battle in the civil war that raged across the land from the year of Newton's birth in 1642. No epitaph-bearing gravestone survives Horrocks, but Crabtree on the back of a bundle of Horrocks's letters wrote, "Thus God puts an end to all worldly affairs. I have lost, alas, my dear Horrocks. Hinc illae lachrimae [hence those tears]."

Many of Horrocks's papers were taken by his brother to Ireland, where they were subsequently lost. It was a historian, Christopher Townley, who some years later collected what remained, and eventually these were studied by Flamsteed, Newton, Wren, Halley, and others of that remarkable circle. What thoughts and results Horrocks's manuscripts may have engendered in others we do not know, but how fascinating to consider what might have been had he lived.

The Mysterious Gamma-Ray Bursters

Chapter 16 celebrated the seventy-fifth anniversary of what has become famous in twentieth-century astronomy as the Great Debate. For reasons perhaps more human than scientific, the debate caught the popular imagination and has been a staple of elementary astronomy texts ever since.

It was the idea of Robert Nemiroff of NASA and George Mason University that the diamond jubilee of the Great Debate should be celebrated by having another such debate, this time over a puzzle that, like its predecessor, has exercised the minds and tongues of astronomers for decades without a firm solution having been found. It is the puzzle of what and where are the sources of brief but violently energetic bursts of gamma rays that come at us every day from somewhere beyond the earth.

The contenders in this new debate (held on 22 April 1995, in the same hall as the first debate) were Donald Lamb of the University of Chicago and Bohdan Paczyński of Princeton University, with Sir Martin Rees of Cambridge University as moderator. Basically, it was Lamb's position that the gamma-ray bursters (as the sources have come to be called) are neutron stars in a giant corona surrounding our galaxy, while Paczyński argued that the bursters, of unknown identity, must lie at cosmological distances some ten thousand times greater than that.

Like so many discoveries in astronomy, the discovery of the gamma-ray bursters was serendipitous. The story begins in the 1960s during the cold war, when the two sides in the war anxiously monitored each other's activities, particularly with regard to nuclear-bomb tests. Fearing the Soviet Union might go so far as to carry out tests on the far

First published in *American Scientist*, 84:434, September–October 1996.

side of the moon, where the explosion would be invisible from earth, the United States put into orbit pairs of satellites that circled the earth opposite one another at a separation of about 150,000 miles and could watch almost the whole sky and surface of the earth continuously. An actual nuclear explosion on the far side of the moon might still not be seen directly, but because the subsequent blast cloud of dust and debris would be radioactive and emit x rays and gamma rays well above the surface of the moon, the satellites were equipped with specially designed detectors of such wavelengths.

The years went by, and of course no nuclear explosions on the moon were ever detected. But the satellites did report occasional bursts of gamma rays coming from random directions at a rate of ten or twenty a year. Timing the moment of arrival of a given burst at various satellites allowed something to be said not only of the direction of the source but also whether it was nearby. It turned out that the sources must lie beyond the inner solar system; the major planets and sun were therefore ruled out.

The gamma-ray reports were not classified by the military, and eventually they came to the attention of astronomers, the first announcement of them being a paper in the *Astrophysical Journal* in 1973. Naturally, although the direction determinations were relatively poor, everyone wanted to know whether the bursts could be identified with one or other of the many strange objects that populate the astronomical zoo. The answer was no; as far as one could tell, they came from no clearly identifiable type of object. An even greater surprise was that reports of gamma-ray bursts were not accompanied by reports of bursts at any other wavelength. So far as we know even today, the sources, whatever they may be, emit bursts only at these most energetic frequencies and at no other, not even x rays.

The situation was a paradise for theorists, untrammeled by facts. Only two years after that first *Astrophysical Journal* paper, an exasperated reviewer remarked that there were already more theories than the total number of bursts reported to that time! And by now, a quarter-century later, more than two thousand papers have been written about gamma-ray bursters, yet, as Gerald Fishman, a veteran of the field, puts it, "they remain perhaps the least understood of all objects that have been observed in the Universe."

The observational situation, however, has improved over the years and brought new constraints to the theorists. As more sensitive detectors have been placed on satellites, the rate of detection has gone up from a dozen or two a year to more like three hundred a year, or nearly one a day. For years some of the best data came from an experiment running on the *Pioneer Venus Orbiter* (PVO), but in April 1991 the Burst and Transient Source Experiment (BATSE) was put into orbit on the *Compton Gamma-Ray Observatory*, and this has almost revolutionized the subject.

With the advent of the PVO data, it seemed clear that the bursters were randomly distributed in the sky, and that their observed intensities followed a so-called three-halves law. That is to say, the cumulative number N having intensities above some level P depended on P to the power of -1.5. This implied that the sources are homogeneously distributed in three-dimensional Euclidean space around us, suggesting we are detecting objects that are relatively local, say within about ten light-years.

Theories based on these results included the possibility that the sources lie in the Oort Cloud of comets on the distant outskirts of the solar system. This spherical shell of perhaps 10^{11} cometary nuclei starts at about fifty thousand times the earth's distance from the sun. How to produce the gamma-ray bursts there is a bit tricky, but comet-comet and comet–primordial-blackhole collisions have been posited. In any case, this theory seems to have fallen from favor. The Oort Cloud cannot be exactly spherical because of tidal effects of our galaxy, and the number density of comets almost certainly must vary with distance.

The most favored theory at the height of the PVO era was that the bursts somehow involve neutron stars on an order of a few thousand light-years away in the thick disk of our galaxy. A neutron star is the remnant core of a moderately massive star that has exploded as a supernova. It is typically about six miles in diameter but with a mass several times that of the sun, so that the star is at nuclear densities—hence its name. Its magnetic field may range to 10^{12} times that of the earth or more, and all in all its exotic properties make a good starting point for theoretical models of gamma-ray bursters.

BATSE changed all that. Some thirty times more sensitive than the PVO detectors, those on BATSE were expected to finally reveal the nonisotropy due to there being a concentration of neutron stars in the galactic disk. If BATSE could "see" the faint sources outside the disk, it should find fewer there, and the sky maps would begin to show a higher density of points in the galactic plane. In fact, since we are located a considerable distance away from the galactic center, it was expected that the direction to the center would show an even heavier concentration of bursters on the sky maps.

To the astonishment of almost all burst watchers, BATSE did nothing of the kind. Certainly it detected many more weaker sources, but the isotropy was only enhanced, and (horror of horrors) the three-halves law broke down: for really weak sources, N varied as P to the power of − 0.8. This meant either that BATSE had penetrated to the edge of the burster distribution, or that it was dealing with a distribution on a scale that required non-Euclidean space, that is, a cosmological scale of perhaps a billion light-years. Where were the theorists to turn?

Here we come back to the debate between Lamb and Paczyński. Lamb favored a theory that still invoked neutron stars as burster sources, but now located them in an immense halo or corona on the order of several hundred thousand light-years' radius around our galaxy, while Paczyński much preferred the billion light-year cosmological scale, even though there is currently no known satisfactory model for the sources themselves at such distances.

Neutron stars in general have very rapid rates of rotation, some measured in milliseconds per rotation. The intense magnetic field of such a star beams the nonthermal electromagnetic radiation it emits (particularly at radio wavelengths), and the rotation then sweeps this beam around just as a lighthouse sweeps its beam of light around. If we happen to lie near the plane of the beam, radio telescopes can detect the train of pulsed radiation as it repeatedly sweeps over us, and such neutron stars are called pulsars. This, in fact, is almost the only way of detecting neutron stars.

Lamb based his preference mainly on two facts: first, that it has recently been discovered that perhaps as many as half of all pulsars have velocities over five hundred miles per second, and in some cases at least twice this value. At such speeds they would move out into the postulated corona in relatively short times of less than a hundred million years, and would appear isotropically distributed. Second, there are three known cases where a gamma-ray burst has been repeated by the same source, and all three are associated with young supernova remnants (the dispersed material of the exploded star), which, of course, are the birthplaces of neutron stars. In particular, one of these three is located in the Large Magellanic Cloud at a known distance of 163,000 light-years, comparable to the postulated corona at several hundred thousand light-years. Among its seventeen recorded bursts is one whose characteristics overlap many of those of the nonrepeating bursters.

Lamb cited other evidence linking bursters to neutron stars, in particular early evidence that some bursts have shown spectral lines, so-called cyclotron lines. These require a strong magnetic field to explain them, and again it is neutron stars that have such fields.

Paczyński, however, discounted much of this evidence. He argued against there being a corona of sources, noting that the Andromeda galaxy, much like our own but two million light-years away, presumably also would have such a corona, yet does not show up at all on sky maps of the bursts. A counterargument would be that BATSE lacks the sensitivity to detect neutron-star bursters at that distance, but Paczyński responds by saying that at their postulated speeds some of the Andromeda bursters would have reached our galaxy by now, yet we see no increase in the direction of Andromeda, so either there are none or they switch themselves off after a certain time. More particularly, however, Paczyński disagreed that sources in such a corona, if they had originated in the galactic disk, would appear isotropic. His adamant assertion was that the signature of their birthplace would always be revealed in their present distribution.

As to the repeating bursters, Paczyński noted that while there may be some overlap in characteristics, those of the repeaters are sufficiently different from the nonrepeaters to suggest a different kind of source. He went on to note that the claims for cyclotron lines were uncertain, and BATSE, the most reliable and sensitive of available instruments, has so far not found any evidence for such lines.

Paczyński's favoring of the cosmological scenario seems mainly a matter of default; his dissatisfaction with the galactic corona theory leaves only the cosmological one as an alternative. The strong point of this theory is that it easily explains the isotropy of sources as well as the breakdown of the three-halves law, which would be due to the expansion of the universe (a redshift effect). There is more positive evidence in that weaker bursts seem to last longer, which could be a relativistic time-dilation effect for more distant sources, but as yet this is not really certain. A significant weakness of the theory is that it offers no clue as to what actually causes bursts so powerful that we see the signal from across the universe.

Of course there was much evidence and discussion in the debate not presented here, but I think it fair to say that, as in the first Great Debate, there was no clear victory for either side, and the moderator wisely did not call for a vote.

What of the future? A critical test will be to check whether or not the Andromeda galaxy has a corona of bursters. If it does, then the corona theory wins; if it doesn't, the cosmological theory wins. It will take detectors one to two orders of magnitude more sensitive than BATSE to check this, but presumably that will come. Meanwhile, new, suitably located satellites will improve direction finding of bursts and testing for counterpart bursts at other wavelengths. Already *Nature* (4 April 1996, p. 377) reports a rapid-response system that flashes a burst detection from BATSE to the Goddard Space Flight Center in Maryland, where the burst coordinates in the sky are instantly calculated and sent across the Internet to a network of some thirty fast-slewing optical and radio telescopes around the world, so that within seconds of a gamma-ray burst, the same area of the sky is being searched at longer wavelengths. So far, though, no luck!

We smile now when looking back at the original Great Debate, knowing as we do what the answer really was, and I'm confident that when the time comes to celebrate the hundredth anniversary of that debate, there will be smiles again in looking back to the 1995 debate.

Transits, Travels, and Tribulations. I

I recommend it therefore again and again to those curious astronomers who (when I am dead) will have an opportunity of observing these things, that they would remember this my admonition, and diligently apply themselves with all their might in making this observation, and I earnestly wish them all imaginable success: in the first place, that they may not by the unseasonable obscurity of a cloudy sky be deprived of this most desirable sight, and then, that having ascertained with more exactness the magnitudes of the planetary orbits, it may redound to their immortal fame and glory.
—*Edmond Halley, 1716*

Inevitably, we take the sun for granted. If it's unclouded, that's good; if it's producing a beautiful sunset, that's even better; its rising and setting govern the timetables and seasons of our lives; every vestige of life on earth depends on its continuing to shine; but as to just what it is or how far away it is, well, if you can't remember what you learned in grade school, you can always look it up on your CD-ROM encyclopedia. People once worshipped the sun as a god, and perhaps our own age could do with a little of the awe it once inspired.

The task of finding out about the sun begins with determining its distance. Once known, this will lead us to the sun's mass (its repository of fuel) and rate of energy output, and so eventually to its lifetime. Not to mention the myriad interesting aspects of its physics. Moreover, the distance of the sun (actually its average distance from earth) defines a basic unit in astronomy, known simply as the astronomical unit (AU). Not only is this convenient for discussions of planetary motions, but it once formed the first step in the long ladder of astro-

First published in *American Scientist*, 85:120, March–April 1997.

nomical distances. Distances of stars were first determined from their trigonometric parallaxes, this being the change in direction of a star as seen from opposite ends of a baseline of length known in AUs, formed by the earth's orbital motion. More extensive methods of finding distances by an application of the inverse-square law of light could then be calibrated from known stellar parallaxes. But how to calibrate the astronomical unit itself?

It took some two thousand years from the first start at this problem before a reasonably accurate value was achieved. The fragmentary record does not show that any of the early Greeks arrived at a specific number of stadia (the stadium being then an everyday unit of length) for the sun's distance, but they might have. Circa 250 B.C.E., Aristarchus of Samos carried out observations that led him to conclude the sun is about nineteen times farther away than the moon. At about the same time, Eratosthenes determined how many stadia there are in the earth's radius, and a couple of generations later Hipparchus found the moon's distance to be sixty-four earth radii. He would likely have known of the earlier results, and by simply multiplying the three figures together could have arrived at a distance for the sun that would have been the equivalent of about four million miles.

This is a far cry from today's figure of about ninety-three million miles, yet there was little advance right up to the time of Kepler in the early 1600s. Kepler himself is usually credited with a figure corresponding to about fourteen million miles, although his Rudolphine tables, published near the end of his life, have been shown to be based on the traditional four million miles value. In any case, even his larger figure was the result of little more than hand waving. Nevertheless, it was Kepler who opened the door to much-improved ways of finding the sun's distance through his discovery that the square of a planet's orbital period is proportional to the cube of its average distance from the sun.

This, the third of his famous laws of planetary motion, enables us to establish the scale of the planetary system in AUs. To take the case of Venus as an example, simple observations of its angular distance from the sun as a function of time lead to the result that its orbital period is 225 days. The earth's orbital period, of course, is 365 days, so that we may use Kepler's third law to write $(225/365)^2 = (d/1)^3$, where d is the average distance of Venus from the sun in AUs, the earth's distance being 1 by definition. This tells us that d = 0.72 AU. Repeating this for all the other planets allows us to construct a map of the solar system on which the relative positions of the planets vis-à-vis the sun are shown on a scale of astronomical units. If now an absolute distance between the earth and another planet at any particular time can be measured, then the ratio of that distance to the corresponding distance on the map will yield the value of the astronomical unit in absolute units—that is, the distance of the sun in miles. For instance, if we can determine that Venus is twenty-six million miles away at its closest approach to earth, then we know that $(1 - 0.72)$ = 0.28 AU corresponds to 26 million miles.

Until our own day, such a distance had to be determined trigonometrically, using observers as widely separated on the earth as possible to observe the direction of the other planet against, say, the backdrop of distant stars. The difference in direction as seen from the ends of the known baseline between the observers would give the planet's distance in miles.

Clearly, the nearer the other planet, the more accurate will be the result. Venus is the planet that comes closest to the earth, but the trouble is that when it is closest, it is more or less between us and the sun and thus is near the sun in the daylight sky, where neither it nor the background stars can be seen. This led early observers to concentrate on Mars, the planet that has the second closest approach to the earth. When it is closest to earth, it is highest in the midnight sky—the most desirable situation. Thus in 1672, Jean Richer was sent from Paris to Cayenne on the northern coast of South America to observe Mars on such an occasion, while Jean Dominique Cassini, the director of the Paris Observatory, stayed home to observe Mars from there. (It was Richer's finding that his carefully calibrated pendulum clocks ran slow at the five degree latitude of Cayenne that later led Newton to claim the earth must be oblate. This was hotly disputed by the French, and eventually resulted in the famous expeditions of La Condamine to Ecuador and de Maupertuis to Lapland, as we saw in chapters 7, 8, and 9.)

Richer and Cassini found that the direction of Mars differed by about 0.007 degree as seen from opposite ends of the Cayenne-Paris baseline. From this Cassini inferred a solar distance of eighty-eight million miles, which we now see to have been a huge improvement

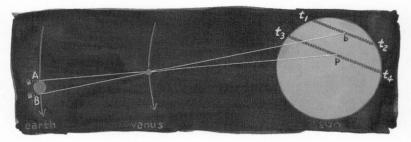

Figure 19.1 *Transit of Venus across the sun as seen by two terrestrial observers, A and B. Each observer sees Venus projected against a different part of the sun's disk, and this difference in position of the two tracks allows the determination of the angle at Venus between A and B. This, together with the baseline between A and B, yields the distance of Venus from the earth. Observer B sees the transit along the upper dashed line, starting at time t_1 and ending at time t_2. Observer A sees Venus transiting along the lower dashed line, starting at time t_3 and ending at time t_4. (Original diagram by Tom Dunne.)*

on earlier estimates, even though it isn't clear what uncertainty he attached to that figure. Possibly the uncertainty can be judged from that fact that when Flamsteed, later Britain's first astronomer royal, included his own observations, he obtained eighty-one million miles, while Picard, an assistant to Cassini, determined 40 million miles from the French data! Whatever the truth, knowledge of the sun's distance was still considered very unsatisfactory when the eighteenth century dawned.

Which brings us back to Kepler, even though he had died seventy years earlier. It was he who had found a way of using Venus after all, making a virtue of the fact that it and the sun are in the sky together just when it is closest to us. With Venus interposed directly between us and the sun, we could expect to see Venus as a black blob slowly transiting the sun's disk. As figure 19.1 illustrates, the transit's track would appear at slightly different positions on the solar disk as seen from widely separated places on earth, from which it would be possible to infer the angle at Venus between the two tracks. That, in turn, equals the angle at Venus between the two observers, A and B. Knowing this and the baseline AB between the observers, one can calculate the distance of Venus from the earth. In short, Kepler had suggested substituting the sun's disk itself as a reference frame in place of the background stars for making the same kind of observation that Richer and Cassini made for Mars. The advantage, as mentioned, lay in Venus being much closer to the earth than Mars, with a corresponding improvement in the accuracy of the final result.

Transits would be seen about every nineteen months if Venus's orbit were in the same plane as the earth's orbit, but unfortunately Venus's orbit is tilted by about 3.4 degrees to that of the earth. The disk of the sun, on the other hand, subtends an angle of only about 0.5 degree as seen from the earth, or about 0.25 degree on each side of the earth's orbital plane. Thus on most occasions when Venus is between us and the sun, its direction is above or below the sun's disk, and we do not see a transit. Only when both earth and Venus are near the line of intersection of their respective orbits, and Venus between earth and sun, will a transit be visible. This greatly reduces the frequency of transits; Kepler himself predicted in 1627 that a transit would happen on 6 December 1631, and then not again for another 130 years. He died the year before this predicted date, but a French philosopher, Pierre Gassendi, was eager to see the transit. Unhappily, "an impetuous storm of wind and rain rendered the face of the heavens invisible," but in any case, as we now know, the transit took place during the European night of 6 December, and would have been invisible to Gassendi.

Only a few years later, Jeremiah Horrocks in England reworked Kepler's calculation and found that another transit was due in 1639, Kepler having thought it would be a near miss. As we saw in chapter 17, Horrocks discovered this only weeks before it was to take place, and did succeed in observing it. But since no one else in distant lands observed it, no calculation of Venus's distance could be made, although the smallness of Venus's disk against the sun suggested that the scale of the solar system must be significantly larger than previously thought.

Thanks to Horrocks, then, it was soon known that in modern times transits of Venus come in pairs separated by eight years. There would be transits in 1761 and 1769, 1874 and 1882, and 2004 and 2012. A combination of the rarity of these events and the importance they held for determining the scale of the solar system led to increasing discussion and preparation as the eighteenth century advanced.

One consideration centered on the most accurate way in which to make the actual observations. In figure 19.1, Venus approaches the disk of the sun from the left, and its transit across the disk as seen by A and B appears as dotted lines. In principle all that is needed is for each observer to measure the angle between the midpoint of the track and, say, the upper edge of the sun. Then the difference between the two measured angles is the needed angle. In reality, however, the separation of the two tracks is only one-fiftieth of the solar diameter, or about 0.01 degree, even for a baseline between observers equal to a full earth diameter. Add to that the fact that the observers would be using relatively small, portable instruments in the field under some-times hostile conditions, and one could not hope for much accuracy in this important and expensive endeavor.

Edmond Halley gave much thought to the eighteenth-century tran-sits, even though he knew there was no hope of his living to see them (he died in 1742, in his eighty-sixth year), and it was he who proposed an indirect method that would be much better. Returning to figure 19.1, it is clear that the length of each track on the sun's disk depends uniquely on how close it is to the center of the disk: the farther from the center of the disk, the shorter the track. Given the length of each track, it would be easy to locate them on the disk and hence their angular separation, the angular diameter of the sun being known. Halley went one step further: a measure of a track's length would be the duration of the transit. In other words, if the observer had a reli-able clock, then all that was needed would be to note t_1, the time the transit started, and t_2, the time it ended, while another, distant ob-server would correspondingly note t_3 and t_4. A transit would take over six hours, and if the times could be judged to within a second or two, a precision on the order of 1:10,000 might be achieved—far better than could be expected from direct angular measures.

A useful variant of this approach was developed by the French astronomer Joseph-Nicolas Delisle. Again, it is clear that the starting time t_3 is earlier than the other starting time t_1, or conversely, the ending time t_4 is later than the corresponding t_2, and it is possible to use the differences $(t_3 - t_1)$ or $(t_4 - t_2)$ to establish the separation of the tracks.

The two methods were neatly complementary. Halley's method had the advantage that one had only to measure the difference in time between start and finish of a transit as seen from a given site, but it required the sky to be clear for both; if one started and five hours later it clouded over, one's efforts were worthless. Conversely, Delisle's method needed a measure of only the start or the end of a transit, not necessarily both. Its disadvantage lay in the fact that the times had to be absolute, for eventually one had to compare the times on two clocks a world apart, and this was a very severe strain on eighteenth-century timekeepers. The choice of sites also depended on which method was to be used: in Halley's method the site had to be such that the sun would be above the horizon throughout the transit, whereas Delisle's method allowed sites where the sun would set or rise in the course of the transit. Both schemes were subsequently used with success.

As the years rolled on toward 1761, preparations became more urgent. Many countries were involved, particularly the two super-powers of the day, Britain and France, and since these two were all too often at war with one another in the eighteenth century, the send-ing of observers to remote places would be dangerous indeed, made even more hazardous by the fact that much of the world was as yet unexplored. It would be a time of high adventure, and some of the participants would become famous. We shall look at the sweep of it all in the next chapter.

Transits, Travels, and Tribulations. II

The record shows that well over a hundred observations of the 1761 transit of Venus were made, many by observers other than French or British, but the large majority of these simply observed from their home stations in Europe. And even the French and British planned only a few expeditions to remote places, essential though they would be for obtaining a useful result. Money unquestionably played a role here; it was an age before mass production, and a single good, portable telescope suited to a transit observation cost some fourteen hundred pounds, which probably exceeded the annual salary of the astronomer royal by a considerable factor. And more than one telescope per expedition was needed in case of accidents. Little wonder that the Royal Society of London, which was responsible for the British expeditions, tried borrowing or renting equipment whenever possible.

Given that the first observation of a transit of Venus had been made by an Englishman, Jeremiah Horrocks, and the importance of such observations emphasized by another, Edmond Halley, the British were surprisingly slow in preparing for the June 1761 transit. Only in the summer of 1760 did the Royal Society get down to details of who would go where, and this in an age when it was hoped that a relatively simple voyage down the Atlantic to the island of St. Helena and back would be completed within a year.

Here, however, the British were helped in their planning by the French. In charming contrast to our own times, we find that while their governments and associated armies and navies ferociously fought one another, the Royal Society and Académie royale des sciences saw no reason not to continue their cordial relations, even if exchanges of

First published in *American Scientist*, 85:418, September–October 1997.

documents required more circuitous routes than usual. So both institutions were aware of each other's plans, and in particular, Joseph-Nicolas Delisle, who had known Halley and been inspired by his zeal for the transits, communicated to his Royal Society counterparts a lengthy analysis of the situation and French intentions. These included expeditions to Siberia, India, and the island of Rodrigues in the Indian Ocean.

To complement the expeditions, it was desirable to have a station in the South Atlantic, so initially the British decided on just one expedition, which would go to St. Helena. It would be led by Nevil Maskelyne, later to become astronomer royal, accompanied by Charles Mason, an experienced observer and assistant to the current astronomer royal. (The British always sent at least two observers on each expedition in case one should die or be otherwise incapacitated. Indeed, in the light of naval warfare, experts advised that the two should "go in *different Ships* . . . to [avoid] the Risque of both being Embarked on the *Same* Bottom." Sound advice, but economically unfeasible.)

Later, however, it was decided to extend the longitudinal range of the sites by having the British send a second expedition to Bencoolen (today's Bengkulu), a small port on the southwest coast of Sumatra. This would be a long voyage, and it being already September 1760, there was no time to waste, so Charles Mason was abruptly switched to being principal observer for Bencoolen, and an assistant for him hurriedly found in the person of a land surveyor and amateur astronomer, by name Jeremiah Dixon. The Royal Navy was petitioned to provide a ship, which would be faster than scheduled sailings of the East India Company's ships and would offer better protection against French attack. This was granted, and by early December 1760, Mason and Dixon were aboard HMS *Seahorse*, and "waited only for the wind."

Only hours out of Portsmouth, however, the *Seahorse* ran into a thirty-four-gun French frigate, *le Grand*. After committing an hour of mayhem on one another, both sides withdrew, and the *Seahorse* limped back to port with eleven dead and thirty-seven wounded. Mason and Dixon reconsidered Bencoolen. Although *Seahorse* was being repaired with all haste, and the navy issued assurances that this time she would be escorted through the English Channel by a seventy-gun man-of-war, the observers' enthusiasm was on the decline. In fact, Mason wrote to the Royal Society firmly advising that "We will not proceed thither, let the Consequence be what it will."

The society's reply minced no words. It warned that "their refusal to proceed upon this Voyage, after having so publickly and notoriously ingaged in it . . . [would] be a reproach to the Nation in general, to the Royal Society in particular, and more Especially and fatally to themselves. . . . [It] cannot fail to bring an indelible Scandal upon their Character, and probably end in their utter Ruin." In case this wasn't clear, the society went on to say that it would "with the most inflexible Resentment" take Mason and Dixon to court and prosecute them "with the utmost Severity of the Law." A succinct reply to this on 3 February 1761 announced that "their dutiful servants" would sail that same evening.

The next communiqué came from South Africa, dated 6 May 1761. In it, Mason triumphantly noted the news that Bencoolen had been taken by the French, and that he and Dixon would be staying right where they were in Cape Town to observe the transit. As it turned out, it was just as well. Maskelyne and his assistant were clouded out on St. Helena, and the successful Mason-Dixon data were all that came from the critically important South Atlantic region. The accuracy of those observations showed the abilities of the team, and so it was that two years later they were chosen to survey the boundary line between Maryland and Pennsylvania, for which their names will forever be inscribed in American history books.

Probably to the surprise of the Royal Society, a third British expedition took place in 1761. This was the work of John Winthrop, professor of mathematics and natural philosophy at Harvard, who petitioned the Province of Massachusetts to provide the means for an expedition to St. John's, Newfoundland. The 1761 transit did not favor the Americas as observational sites since much of the transit took place during their night, but at the northeastern location of Newfoundland the egress of Venus would be potentially visible. The governor, appealing to the House of Representatives that this was a "phenomenon which has been observed but once before since the Creation of the World," was successful in acquiring the province's sloop for the voyage, Harvard provided the necessary observational equipment, and Winthrop and his three assistants sailed from Boston on 9 May 1761. They arrived at St. John's two weeks later and, given every assistance necessary by the local authorities, began practicing observational procedures "with an assiduity which the infinite swarms of insects, that were in possession of the hill, were not able to abate, tho' they persecuted us severely and without intermission . . . with their venemous stings." Happily, the sunrise of 6 June was clear and calm and allowed five precise measures of Venus's path across the sun as well as timing the actual moment of egress.

By and large, things did not go as well for the French expeditions. Alexandre-Gui Pingré left Paris on 17 November 1760 for his destination of the island of Rodrigues, viewing his forthcoming voyage with foreboding. This despite another remarkable novelty of the times. Although Britain and France were locked in a bitter battle, the Académie royale had appealed directly to British authorities to grant Pingré a laissez-passer or letter instructing all British naval and military personnel "not to molest his person or Effects upon any account, but to suffer him to proceed without delay or Interruption." This was indeed granted, although since participants in sea battles tended to exchange gunfire first and civilities later, if at all, Pingré's misgivings were not misplaced.

The transit party sailed on the *Comte d'Argenson*, a warship that found itself with less than half its normal complement of guns in order to extend its cargo capacity to that needed for the expedition. (There had been a heated dockside argument over the baggage, Pingré arguing furiously that seven or eight hundred pounds was not too much for an astronomer!) To the horror of all on board, a group of five British warships was sighted only one day out of port. To allow full play of its remaining guns, the ship's crew tore down the temporary cabins that had been erected for Pingré's companions, the latter and their belongings being flung unceremoniously into Pingré's cabin for the time being. Fortunately, though, a combination of suitable winds, the long winter night, and the captain's skills allowed the *Comte* to slip away unmolested, and everyone settled down to the remaining four months of their voyage.

Methods for determining longitude at sea in the mid–eighteenth century were so poor that an interesting though friendly rivalry sprang up between Pingré and the ship's officers as to their location. Each would determine longitude by his preferred method, and then compare notes. Thus it was that in late January the ship was bearing down fast on the Cape Verde Islands off West Africa, with Pingré announcing that the ship would pass to the east of the islands, while the navigator claimed passage would be to the west of them. The captain hove to for the night in case the average proved correct. The next morning they sighted the island of Santiago some leagues to the west, although according to their charts and the navigator's latest calculation they should have been several leagues inland on the island of Bonavista, which, fortunately, was nowhere to be seen.

They rounded the southern tip of Africa in April, just two weeks before Mason and Dixon would arrive there, and were looking forward to a comfortable run up the Indian Ocean to Rodrigues when they encountered another French ship, *le Lys*. Everyone was delighted, especially since the *Lys* carried fresh supplies of Cape fruit and wine, but delight turned to despair when it turned out that the *Lys* had been severely damaged in an encounter with British ships, and that its captain, superior in rank to the captain of the *Comte*, was demanding that the latter escort the *Lys* at its limping pace to Isle de France (today Mauritius) for repairs. This meant that Pingré would not reach Rodrigues in time for the transit. He was outraged, and a violent exchange of notes ensued. But within a few days he and the captains dined together, digestion being aided by a remarkable flow of wine, and a compromise was reached. The *Comte* would sail ahead to Isle de France.

This ended with the astronomers having to transfer to another ship at Isle de France for the final three-hundred-mile run to Rodrigues. Bad weather prolonged the voyage for nineteen days, and then, with the transit only ten days away, their ship was becalmed for two days within sight of the island. Once ashore, the astronomers rushed to offload their equipment, find a suitable site for observing the transit, and set everything up. The instruments were in poor condition after months at sea, but after frantic effort, everything was ready and tested by the evening of 5 June, with the transit due to start the next morning and continue through much of the day.

The observers awoke to a steady rain. The moment of the transit's start passed without observations being possible, but later the sky cleared somewhat, and Pingré and his assistant got some measures through passing clouds of Venus's position on the solar disk. They hoped to time the moment of egress, but a solid overcast came back and ruined any such hope. Still, their observations were useful, and, astronomy aside, they had been collecting local flora and fauna to take back to the Académie royale. This continued until, at the end of June, a major catastrophe struck with the sudden arrival of a British man-of-war. Its crew captured one of the two French ships in the harbor at the time, burned the other, and ransacked the island. Pingré, of course, rushed forward with his letter from the British Admiralty ordering that his expedition not be molested, only to have the British captain ignore it entirely. When the British finally departed, there was little left for Pingré to take home, and indeed he and his companions were reduced to living on rice and flour and "ignoble de l'eau" while they waited for another French ship to come by on its regular run from Isle de France to pick them up. Pingré protested vehemently about his treatment in a whole series of letters to the British Admiralty and the Royal Society, but it isn't clear that he ever received any satisfaction.

It was not until the end of May 1762, a year after the transit, that Pingré and his companions once more set foot in Paris, exhausted but happy. And although Pingré was never part of the 1769 transit expeditions, he played a major role in planning them and eventually analyzing the results of both transits. Astronomer, theologian, classicist, geographer, traveler, Pingré has been described as "one of the most interesting men ever to enter the ranks [of the Académie royale]." It seems fitting that he lived to a fine old age, dying in Paris in 1796 at eighty-four.

There were two other French expeditions to observe the 1761 transit, one to Siberia and another to India, both, like Pingré's, filled with adventure. We shall take them up in the next chapter.

21

Transits, Travels, and Tribulations. III

There were two other French expeditions to observe the 1761 transit of Venus, that of Jean Chappe d'Auteroche to Siberia, and that of Guillaume-Joseph-Hyacinthe-Jean-Baptiste Le Gentil de la Galaisière to India.

Chappe came of a family from the lower French nobility, but since he was only thirty-one when he was admitted to the Académie royale des sciences in 1759, and since the 1769 transit cost him his life, his career was short-lived and we know little about him. Certainly his early entrée to the Académie royale and the work he did on his transit expeditions bespeak a person of talent and determination, and who knows what he might have achieved in a longer life?

Through an invitation from the Russian Imperial Academy of Science, the Académie royale appointed Chappe to observe the transit of 6 June 1761 from Tobolsk, a city in central Siberia some three thousand miles from Paris. This site was chosen because from it both start and finish of the transit would be visible, granted clear weather. Protocol dictated that Chappe pay his respects to the Russian Academy in St. Petersburg en route, and since travel would be slow, he would necessarily face a crossing of the Ural Mountains and Siberian travel in winter.

Chappe initially hoped to reach St. Petersburg, the halfway mark of his journey, by ship, traveling around the coasts of northwestern Europe. He further hoped to avoid the wretched war between France and Britain by booking passage on a Dutch ship, but a delay in organizing "un appareil considérable d'Instruments" made him miss the sailing. He later drily noted his consolation that this probably saved his life when the ship ran aground on the coast of Sweden. Travel by

First published in *American Scientist*, 86:123, March–April 1998.

land it would have to be, aiming for Strasbourg, Vienna, Warsaw, St. Petersburg, and Moscow, and then striking out across the Siberian plains and over the Urals for Tobolsk. Chappe's expedition left Paris in late November 1760.

It says something of the times that even within civilized France the journey to Strasbourg, a matter of hours in a car today, took eight days over highways so bad that every thermometer and barometer was broken and the carriages damaged beyond repair. What must Chappe have thought, contemplating that this was only the first small step of his journey and that conditions to come would be worse— much worse? But again in tune with his times, he simply set himself to making a new set of instruments while new carriages were arranged. He did, however, decide to head first for Ulm, and thence go by boat down the Danube to Vienna. He was strongly advised against this, since it was the season when heavy river fogs could delay boats for days, but he took to the river nevertheless. Again one is struck by the "Renaissance man" nature of these eighteenth-century scientific expeditionaries: no lounging around the boat for Chappe; he was busily mapping every turn of the river, since France lacked such maps of the upper Danube, and when fog on the river left the boat immobile, Chappe was off climbing the surrounding mountains, barometer in hand, to determine their altitudes.

Vienna was reached on the last day of 1760, and after a reception by Maria Theresa, archduchess of Austria, and her husband, Francis, the Holy Roman Emperor, as well as meetings with local astronomers (comparing barometers, magnetic compasses, etc.), Chappe left Vienna on 8 January 1761. It was a cold day, with temperatures around minus nine degrees Fahrenheit, and soon Chappe and his men were having to smash their way by foot through half-frozen river crossings. By 22 January they were in Warsaw, where Chappe heard he was awaited in St. Petersburg "with great impatience." Crossing the frozen Vistula, the expedition for the first time transferred to sleds, with Chappe reporting on "the ease of traveling with sledges; we went on with the greatest velocity."

In St. Petersburg he learned that in the light of his delays, the Russian Academy had given up hope of his reaching Tobolsk in time and had sent out expeditions of its own (which seem never to have been heard from) to nearer sites, but Chappe was determined to reach

Tobolsk before the transit. Thanks to the empress, his expedition was equipped with every necessity, from bread to interpreters, and it left St. Petersburg in early March on four enclosed sleds, each drawn by five horses running abreast. (One pictures a scene out of *Dr. Zhivago* or something from Tolstoy.) The sleds were smashed beyond repair by the time they reached Moscow, but on 17 March they left that city with new sleds, still with nearly 800 leagues (2,400 miles) to go, including the crossing of the Urals. Paradoxically, while they cursed the bitter cold, they prayed for continued cold weather. A thaw would strand them in the Siberian bogs, from which they might never emerge. It took a month (Chappe furiously writing reports on everything he encountered), even though "the surface of the Volga was as smooth as glass . . . and the sledges went on with inconceivable swiftness." His retinue chose to desert him in the depths of a Siberian forest, and Chappe had to hunt them down, pistol in hand, but eventually the expedition found itself in Tobolsk in mid-April, well before the June transit.

Aided by a military party appointed by the local governor, Chappe soon built a working observatory on a nearby mountain and began observations to determine his longitude and latitude, essential for the eventual calculations of Venus's distance. However, Tobolsk lies at the confluence of the Irtysh and Tobol Rivers, and the thaw that set in with his arrival was unusually rapid, with heavy flooding of the town. To the minds of some locals, this was no doubt due to the activities of the foreigner said to be messing with the sun, and mutterings of mob action to deal with him necessitated an increase in Chappe's military guard.

Chappe could sleep only fitfully the night before the transit, even though he reported that "the perfect stillness of the universe completed my satisfaction and added to the serenity of my mind." The day itself proved perfect, and Chappe observed the entire transit. At the start, he says, "I was seized with a universal shivering," but as the hours wore on and success became ever more imminent, "I truly enjoyed [the pleasure of] my observation, and was delighted with the hopes of its being still useful to posterity, when I had quitted this life." Indeed his observations were still prominent in calculating the scale of the solar system more than a hundred years later.

Couriers bearing the essential observational data were quickly dispatched to Paris and St. Petersburg, but Chappe himself stayed on, making further latitude/longitude observations, not to mention notes on everything that came his way. He eventually made a leisurely return trip through southern Russia, arriving back in Paris almost eighteen months after the transit. The only sour note came later, when Chappe published his no doubt honest yet excoriating views on Russian backwardness, despite the help he had received from Russians throughout his travels. No less a personage than Catherine II, writing under a pseudonym, undertook a line-by-line rebuttal. No matter. For the next transit, Chappe was to abandon the winter wastelands of Russia for the deserts of Mexico.

The abbé Le Gentil was another who would pursue both transits. Indeed, his eleven-year odyssey would rank as the longest astronomical expedition in history. Despite a contemporary saying of him that "his face did not prejudice one in his favor," at twenty-eight Le Gentil was a well-trained astronomer. By 1753, he had made his own calculations of the transits and had volunteered to go to Pondichery in India to observe the 6 June 1761 event. He sailed from Brest in a French man-of-war in March 1760, allowing himself plenty of time to sail around southern Africa and across the Indian Ocean. Most of this trip was "uneventful, save for the loss of a fellow passenger by suicide and by the pursuit by an English fleet off the Cape of Good Hope."

When he arrived at Isle de France in July 1760, however, he learned that Pondichery was besieged by the British, and that a French force sent by sea to raise the siege had been all but destroyed by a hurricane while en route. A second force arrived after a delay of eight months, and Le Gentil accompanied it on its attempt to relieve the beleaguered Pondichery. Winds were contrary, however, ("we wandered around for five weeks in the seas of Africa"), and by the time the fleet arrived off the Malabar Coast only two weeks before the transit, the British had captured and consolidated themselves in Pondichery. Le Gentil's ship was lucky to elude the British naval squadrons, and it immediately headed back the three thousand miles or so to Isle de France. In midocean on 6 June, under a cloudless sky, Le Gentil had a perfect view of Venus transiting across the sun's disk. But since precise timing of the event was essential and his pendulum clocks were useless at sea, the view was of no scientific value at all.

Rather than go home empty-handed, Le Gentil wrote to the Académie royale in Paris suggesting that he spend a year or so exploring the islands of the Indian Ocean, carrying out work in natural history, geography, navigation, and more or less anything that might be useful. This was agreed to, and soon Le Gentil was busy mapping the east coast of Madagascar. Here he made the mistake of eating the local beef, which though "rich" caused "a sort of violent stroke, of which several copious blood-lettings made immediately on my arm and my foot, and emetic administered twelve hours afterwards, rid me quite quickly." The accompanying double vision took somewhat longer.

Time went by. The war with Britain ended. The 1769 transit began to loom, and before long Le Gentil was suggesting that he stay on and try again for observations from this part of the world. His latest calculations suggested that going to Manila in the Philippines would be preferable to Pondichery (now back in French hands), and he even considered heading for the Mariana Islands in the Pacific until he learned that ships went there only every three years. Not waiting for agreement from Paris, he seized the chance of passage on a Spanish ship bound for Manila, and it was not until July 1767, a year later, that a reply from the Académie royale caught up with him there, telling him that he should go to Pondichery after all.

As it happened, Le Gentil was encountering considerable hostility from the corrupt governor of Manila (soon to be jailed himself), who apparently disliked all French citizens on principle. He claimed Le Gentil's papers must be false, and Le Gentil sensed that if he did not soon get away, he would probably find himself in a Spanish jail, if not worse. Clandestinely, he left on a ship bound for Madras in February 1768. It was a nightmare voyage, navigating through the islands and straits of the South China Sea, where the captain and his two pilots argued interminably over which passage to head for. Their arguments were so violent that at times all three would storm off to their cabins and leave the helmsman to his own devices. The captain, noted Le Gentil, "was as little in condition to conduct his vessel as I am to lead an army." On the other hand, the pilots were "two old automatons to whom I would not have entrusted the conduct of a launch." Nevertheless, Le Gentil finally found himself in Pondichery on 27 March 1768, more than a year ahead of the transit.

Here he was warmly welcomed by the French governor, Monsieur Law, and an observatory was established for him amid the ruins of a once-palatial estate. In the recent war it had served as a gunpowder magazine, and in fact Le Gentil's observatory was built atop a vault containing "sixty thousand weight of powder." Even so, "this circumstance did not interrupt the course of my observation," said Le Gentil, announcing his pleasure at living and working there. Even the British sent over an excellent telescope from Madras in case it was needed. Le Gentil settled into regular astronomical work, in particular the all-important determination of his precise longitude and latitude.

The year 1768 gave way to 1769, and the transit date of 3 June approached. At Pondichery only the egress of Venus from the sun would be visible, but the precise timing of that event near seven o' clock in the morning local time on Sunday, 4 June, would be vital. "During the whole month of May, until the 3rd of June, the mornings were very beautiful." The evening before was clear and calm. But at two in the morning, Le Gentil was awakened by "the moaning of the sandbar," implying a change in the wind. Leaping from his bed he "saw with the greatest astonishment that the sky was covered everywhere. . . . From that moment on I felt doomed, I threw myself on my bed without being able to close my eyes." A powerful wind brought even heavier cloud, "the sea was white with foam, and the air darkened by sand and dust." Nothing in the sky was visible at seven, but around nine the sun came out, and "we did not cease to see it all the rest of the day."

Le Gentil's journal entry says it all: "That is the fate which often awaits astronomers. I had gone more than ten thousand leagues [thirty thousand miles]; it seemed that I had crossed such a great expanse of seas, exiling myself from my native land, only to be the spectator of a fatal cloud. . . . I was more than two weeks in singular dejection and almost did not have the courage to take up my pen to continue my journal." Especially when he later learned that near-perfect conditions had prevailed in Manila.

His one thought now was to return home, and he arranged passage on the first available French ship, due to leave Pondichery in October. But he fell seriously ill with fever and dysentery, and missed the sailing. Indeed, he barely survived, but by March 1770, he was so desperate for home that, although still ill, he took ship for Isle de

France as a first step. Here his convalescence continued for seven months, until in November he left on a ship bound for home via the Cape of Good Hope. Only two weeks out, though, an extremely violent storm almost sank them, and only great good luck brought them back to Isle de France on New Year's Day 1771. Another three months passed ("The sight of [Isle de France] had become unbearable to me"), but in March 1771, Le Gentil was aboard a Spanish warship, which finally returned him to Europe. On 8 October 1771, "at last I set foot on France at nine o'clock in the morning, after eleven years, six months, and thirteen days of absence."

Le Gentil discovered that since no one in Paris had heard of him for so long, he had been presumed dead, and his seat in the Académie royale had been given to someone else, while his heirs were engaged in dividing up his estate. The latter problem took much expensive and tiresome litigation to correct, but intervention by the king gave him back a seat in the Académie royale within a few months.

There was a happy ending to his life. He lived another twenty-one years, married happily, had a daughter who became the delight of his life, and died of a relatively mild sickness at the age of sixty-seven in 1792.

22

Transits, Travels, and Tribulations. IV

Having looked at the adventures of the 1761 expeditions to observe the first eighteenth-century transit of Venus, as well as the horrors of Le Gentil's ill-fated travels for both transits in that century, we here consider two of the other 1769 expeditions. We saw in the last chapter that Jean Chappe d'Auteroche succeeded in observing the 1761 transit from the town of Tobolsk in Siberia. He volunteered to observe the 1769 transit, asking only relief from snow and ice. The Académie royale des sciences in Paris obligingly sent him to Baja California.

Chappe's entourage was slight. Besides Chappe himself, it included an artist, a clockmaker, and a man named Pauly, described as an engineer-geographer. Two Spanish assistants were added for political reasons, and the party sailed from Cadiz on 8 March 1768, hoping to reach its assigned site before the transit some fifteen months later. Getting across the Atlantic took seventy-seven days, and after landing at Mexico's east-coast port of Vera Cruz, the group set off northwestward to cross six hundred miles of wild terrain. Pauly, who would be the sole survivor of the expedition, writes of "high mountains, dreadful precipices, dry deserts. . . . We came near dying a thousand times. We were besides crushed by an excessive heat which hardly left us strength enough to drag ourselves around," especially after it ruined their food supply and left them to live off the countryside.

Two months of this brought them to "San Blas, on the rosy sea," about 120 miles south of Mazatlán. There followed six weeks of navigating three hundred miles of the Gulf of California to reach San José del Cabo at the southern tip of Baja California ("so dangerous that nobody ever dared to land thither . . . because of perpetual waves

First published in *American Scientist*, 86:422, September–October 1998.

foaming with rage against rocks"). Despite the foaming waves (one sometimes suspects Pauly of laying it on a bit thick), they got themselves and their equipment ashore with only thirteen days to go before the transit on 3 June 1769. That was the good news. The bad news came when "some savage people informed us that a most dreadful epidemic was laying waste to the country." The expedition was advised to move immediately at least a hundred leagues (280 miles) to the north, but they had yet to unpack, clean, and set up their instruments; there was no time for more travel. They stayed where they were.

Chappe's observations of the transit were among the best made of either transit anywhere in the world. Moreover, his subsidiary astronomical observations to establish the latitude and longitude of his site (crucial to the final analysis) were done with an accuracy unprecedented in arduous fieldwork.

But already the epidemic had the expedition in its grip. Pauly writes, "We used to feel the most unspeakable pains, and every one of us . . . was wishing most anxiously for death as a supreme cure. You would hear all around but heavy groanings; every day used to carry away some of our Companions." Chappe himself died on 1 August. "We were all dying, myself and my companions, when I closed up his eyes. . . . Our situation did not allow us to attend to his funeral with many ceremonies."

Pauly, now leader by default of those left, packed up "all the papers concerning the object of our voyage . . . in a casket which I directed to the Viceroy of Mexico. I begged most earnestly some savages of good standing . . . to see that it would reach its place in the case we should all pass away, and to tell the Viceroy to have it shipped [to the academy in Paris]."

Pauly, clutching the casket, and the remaining two expedition members set off on the fearful journey homeward in September. These other two both died en route, and Pauly himself had to pause for strength, sometimes months at a time, before finally arriving in Paris a year later on 5 September 1770. "I hasted to the Academy the observations made in California. That body of men has bestowed on me the highest Eulogy." The king, Louis XV, awarded Pauly a pension of eight hundred francs a year. Sadly, though, it seems he remained an invalid, and eventually had to petition for an increased pension.

The Royal Society of London organized two expeditions for the 1769 transit. One of these had as its leader William Wales, at one time a computing assistant at the Royal Observatory in Greenwich. Aware of Chappe's horrific experiences in the Siberian winter preceding the 1761 transit, Wales requested a warm and not-too-distant site. The Royal Society sent him to Fort Prince of Wales, a Hudson's Bay Company fur-trading post in northern Canada. Even today, as the town of Churchill, Manitoba (population 1,143), on the coast of Hudson Bay, it has no road link to the rest of Canada and is famous for its polar bears. As to payment, Wales and his assistant, Joseph Dymond, would be paid two hundred pounds if, upon their return, their work was judged well done.

Still, the journey was to be straightforward. There were regular sailings of Hudson's Bay Company ships to and from the trading post, so there would be no unpleasant overland treks for Wales and Dymond. On the other hand, pack ice limited the shipping season to two months at Fort Prince of Wales, and it would begin only after the 3 June transit. Thus they would have to be there in the late summer of 1768 and winter over. They sailed from England on 23 June 1768.

By 18 July, they were crossing the Labrador Sea and entering the Hudson Strait, well north of sixty degrees latitude. On the nineteenth, they "passed within a cable's length of a very large island of ice . . . [it] was adorned with spires; and indented in the most romantic manner that can be imagined." Navigating the islands of the straits, they had their first meeting with the Inuit people when "Eskimaux in their canoes, or, as they term them, Kiacks" came alongside. Wales was intrigued by their clothing, especially that of the women, which included "boots [that] come up quite to their hips, which are there very wide, and made to stand off from their hips with a strong bow of whalebone, for the convenience of putting their children in. I saw one woman with a child in each boot top."

On 27 July, "This evening I told 58 islands of ice." Two days later, the ship was among the remnant floes of the pack ice. "It is really very curious to see a ship working amongst ice. Every man on board has his place assigned him; and the captain takes his in the most convenient one for seeing when the ship approaches very near the piece of ice which is directly a-head of her, which he has no sooner announced, but the ship is moving in a quite contrary direction, whereby it avoids striking the piece of ice [directly]. . . . In this manner they turned the ship several times in a minute; the wind blowing a strong gale all the time."

And then there were the weird atmospheric effects of the high Arctic. On 7 August, they caught their first glimpse of Cape Churchill, and yet "though we saw the land extreamly plain from off the quarter deck, and, as it were, lifted up in the haze; yet the man at the mast head declared he could see nothing of it. This appeared so extraordinary to me, that I went to the main-top-mast-head myself to be satisfied of the truth thereof; and though I could see it very plain both before I went up, and after I came down, yet could I see nothing like the appearance of land when I was there."

Happily, though, on 9 August the ship worked its way up the estuary of the Churchill River, and at two o'clock "she was safe moored." The local officials "behaved to us with great civility," and soon the visitors were exploring their surroundings. "Mr. Fowler was so kind as to walk with us about ten miles up the country, which, as far as it went, was nothing but banks of loose gravel, bare rocks, or marshes. . . . Our errand was, to see if we could not find some land likely to produce corn; and in all that extent we did not find one acre, which, in my opinion, was likely to do it." What they did soon find, however, were "three very troublesome insects. The first is the moschetto, too common in all parts of America . . . to need describing here. The second is a very small flie. . . . These in a hot calm day are intolerably troublesome: there are constantly millions of them about one's face and eyes, so that it is impossible either to speak, breathe, or look, without having one's mouth, nose, or eyes full of them . . . . The third insect is much like the large flesh-flie in England; but, at least three times as large: these, from what part ever they fix their teeth, are sure to carry a piece away with them, an instance of which I have frequently seen and experienced."

Also, of course, there was very little by way of building materials. This they had anticipated by bringing with them a complete observatory and cabin for living quarters, which they now hastened to reassemble, "having hitherto had no where to lie but on the floor." They set up and tested their observing equipment, and made some of the necessary auxiliary astronomical observations.

The hot calm days and insects soon gave way to intimations of winter.

Two inches of snow fell on 9 September, and by mid-October "we began to put on our winter rigging; the principle part of which was our toggy, made of beaver skins: in making of which, the person's shape, who is to wear it, is no farther consulted, than that it may be wide enough, and so long that it may reach nearly to his feet."

On 6 November, "the river, which is very rapid, and about a mile over at its mouth, was frozen fast over from side to side, so that the people walked across it. . . . Also the same morning, a half-pint glass of British brandy was frozen solid in the Observatory.

"In the month of January, 1769, the cold began to be extremely intense: even in our little cabbin . . . in which we constantly kept a very large fire. . . . The head of my bed-place, for want of knowing better, went against one of the outside walls of the house; and notwithstanding they were of stone, near three feet thick, and lined with inch boards, supported at least three inches from the walls, my bedding was frozen to the boards every morning; and before the end of February, these boards were covered with ice almost half as thick as themselves." Their sleep was constantly interrupted "by the cracking of the beams in the house, which were rent by the prodigious expansive power of the frost. But those are nothing to what we frequently hear from the rocks along the coast; these often bursting with a report equal to that of many heavy artillery fired together, and the splinters thrown to an amazing distance."

March finally brought relief from the cold, and by the end of April, "the ground was in many places bare," and the hunters started preparing for the spring goose season. Wales and Dymond turned their attention to final preparations for the coming transit of Venus on 3 June.

One would guess that the chances of clear weather at that site on that date would not have been much better than fifty-fifty, but as it happened, all went well. There were cloudy intervals during the six-hour transit, but at the crucial times of ingress and egress "the Sun's limbs [were] extreamly well defin'd," and the astronomers' observations proved a great success.

The ice in the river finally broke up on 16 June, allowing a spate of salmon fishing ("I have known upwards of 90 catched in one tide"), and Wales and Dymond enjoyed a generally relaxed summer while awaiting the supply ship due in late August. They sailed in early September, and after a relatively calm voyage, were back in London on 19 October. Wales, however, was greatly angered when a British customs officer confiscated his much-prized toggy, moose-skin shoes, and other items of Hudson's Bay Company winter clothing. As always, engaging in a violent argument with the customs officer did no good at all.

Little is known of Dymond, and not much more about Wales, but Wales did later accompany Captain James Cook on his second and third voyages to the Pacific. He ended his career teaching mathematics at Christ's Hospital School in London. Among his students were Samuel Taylor Coleridge and Charles Lamb, one of whom remembered Wales as "a good man of plain, simple manners, with a large person and a benign countenance." There have been worse epitaphs.

23

Transits, Travels, and Tribulations. V

This chapter brings to a close the story of the eighteenth-century transits of Venus and the often amazing expeditions to the ends of the earth that they engendered. Before we turn to the ultimate results of these undertakings, we must look at one more of the expeditions, the most famous of them all, the British expedition to the South Pacific for the 1769 transit.

Early analysis of the 1761-transit observations was not entirely satisfactory, and it was expected that the 1769 transit (the last for more than a century) would offer better results. By 1765, Thomas Hornsby, Savilian Professor of Astronomy at Oxford, was urging the European powers to prepare their expeditions: "Posterity must reflect with infinite regret their negligence or remissness; because the loss cannot be repaired by the united efforts of industry, genius, or power." Calculation showed that the South Pacific, as yet hardly explored by Europeans, would be a desirable station, and in case science should not prove attraction enough, Hornsby noted that it would be a "worthy object of attention to a commercial nation to make a settlement in the great Pacific Ocean." Thus it was that the British expedition to the Pacific would have far more hopes behind it than merely establishing the scale of the solar system. Commerce, politics, and empire were not to be denied. The Royal Society of London's estimate that four thousand pounds would be needed to mount the expedition met with little argument, and an appeal to the thirty-year-old King George III was launched. "The Memorialists, attentive to the true end for which they were founded by Your Majesty's Royal Predecessor . . . conceived it to be their duty to lay their sentiments before Your Majesty with all

First published in *American Scientist*, 87:119, March–April 1999.

humility, and submit the same to Your Majesty's Royal Consideration." Royal Consideration quickly arrived at acquiescence.

The society had among its fellows just the man to command such an expedition: Alexander Dalrymple, a former professional sailor with much experience in Eastern seas, and an adept geographer and navigator. But where to find a ship? Clearly the Royal Navy must be the answer, as it had been for Mason and Dixon years before. And then a major snag. The Admiralty, it seemed, had never forgotten the last time it had allowed an astronomer, Edmond Halley, to command one of its ships on a scientific expedition (see chapter 3). The result had been mutiny and the near loss of the ship. The first lord of the Admiralty, Sir Edward Hawke, rather extravagantly announced he would sooner suffer his right hand to be cut off than sign another such commission. So Dalrymple was out. The Admiralty would find its own man. A junior officer, then doing marine-survey work on the St. Lawrence River in Quebec, was picked. His name was James Cook, the ship he was to command, the *Endeavour*.

The next question was just where in the South Pacific the expedition should go. Such reports as existed of islands in the vast ocean were not entirely reliable as to latitude and longitude, and one would not, like Le Gentil, want to find oneself at sea when the crucial moment arrived. But even as the *Endeavour* was being fitted out, there arrived back from the Pacific the good ship *Dolphin*. And what news! It had found an island that was a virtual paradise on earth, an island "such as dreams and enchantments are made of." An island where not only the surroundings were paradisiacal, but the local culture was utterly different from that of Europe. In particular, the sailors had discovered, no doubt within minutes of arrival, that the women were extraordinarily free with their sexual favors. The gift of anything metallic would hasten proceedings even further. The captain of the *Dolphin* had feared his ship would sink at her moorings as her crew enthusiastically ripped the nails from her decks. Her navigators had taken particular care in determining the island's latitude and longitude. Its name was Tahiti. The *Endeavour* would sail for Tahiti. Considerable quantities of nails would be among her cargo.

So on 26 August 1768, the *Endeavour* sailed from Plymouth, bearing southwest for Rio, then round the horrors of Cape Horn and across some forty-six hundred miles of the Pacific to Tahiti, arriving with almost two months in hand before the transit. Joseph Banks (twenty-six years old, later Sir Joseph, and eventually one of the Royal Society's most colorful presidents), who had joined the expedition as scientific leader and botanist, found previous reports to be accurate. "Soon after my arrival at the tent 3 hansome girls came off in a canoe to see us . . . and with very little perswasion agreed to send away their carriage and sleep in [the] tent, a proof of confidence which I have not before met with upon so short an acquaintance."

But cultural differences went well beyond sexual mores. Ownership seemed a very fuzzy concept, and casually stolen goods became a sore point. Particularly when an important astronomical instrument disappeared and had to be hunted down at gunpoint. The English crew set a poor example, as Banks noted in his journal after a near-perfect observation of the transit: "We also heard the melancholy news that a large part of our stock of Nails had been purloind by some of the ships company during the time of the Observation. . . . This loss is of a very serious nature as these nails if circulated by the people among the Indians will much lessen the value of Iron."

The transit observations concluded, Cook, as per instructions, set off southwestward in search of the great southern continent postulated by philosophers of the day as the counterpart to the great land masses of the Northern Hemisphere. Instead, he discovered New Zealand, and spent six months charting its coasts. Setting off westward once more, he ran into the east coast of Australia and then worked northward, charting two thousand miles of coast as he went. That took them into the channel between the coast and over a thousand miles of the Great Barrier Reef. Despite the crew's desperately careful sailing, the beautiful but treacherous reef claimed the *Endeavour*, and although they eventually got her off, they had to beach her for many weeks on the desolate Queensland coast to make repairs.

With supplies running low, the *Endeavour* put in to Batavia (Jakarta) for refreshment and more permanent repairs. So far the crew's health had been fine; indeed, Cook, with his insistence on sauerkraut as a defense against scurvy, was famous for protecting the well-being of his crews, but he had no defense against the malaria and dysentery ("the bloody flux") of tropical Batavia. By the time the ship set off to cross the Indian Ocean, round southern Africa, and sail the length of the Atlantic, nearly half the crew had died, and most of the remainder were severely stricken. But finally, on 13 July 1771, more than two years after the transit, the survivors, weak and shaken, arrived home. Among those they left behind was Charles Green, the expedition's official astronomer. It was reported that he "had been ill some time . . . [and] in a fit of phrensy got up in the night and put his legs out of the portholes, which was the occasion of his death."

It says something of Cook the man that he would undertake two more expeditions to the Pacific despite these experiences. It was, of course, to cost him his life.

So another chapter in the history of the transits of Venus was closed. No one alive then would see another. It remained only to determine how well they had done in arriving at their goal of calibrating the astronomical unit, the distance between the earth and the sun. Three problems hindered the analysis: first, the curious and unexpected phenomenon called "the black drop"; second, uncertainties in the distances between observers; third, the problem of how to combine redundant observations.

The black-drop problem surprised observers. They were trying to determine the exact moment when the edge of Venus's disk just coincided with the edge of the sun's disk as Venus began or ended its transit, but what they saw was an elongated black ligament joining the two edges and persisting even when Venus's disk was clearly within that of the sun. This so surprised and unsettled the observers that even when two of them were standing alongside each other, their reported times could be half a minute apart, when they were expecting agreement to within a few seconds. As Cook himself reported, "This day prov'd as favourable to our purpose as we could wish, not a Clowd was to be seen the whole day and the Air was perfectly clear . . . [yet] we very distinctly saw a . . . dusky shade round the body of the Planet which very much disturbed the times of the Contacts. . . . We differ'd from one another in observeing the times of the contacts much more than could be expected." Even today we are uncertain as to the cause of this phenomenon, but it certainly degraded the timing of the transits.

The accuracy of the final results also depended directly on knowing the lengths of the baselines between observers, in effect knowing accurately the latitude and longitude of each observer. But since methods for determining longitude in the 1760s were inadequate, to say the least, these baselines were not well determined, and the accuracy of the final results was correspondingly diminished.

The third problem reminds us that in addition to the expeditions described here, there were many other observers who viewed the transits from home, if home happened to be in the right hemisphere at the right time. The initial analysts of the data faced the problem of getting the best single answer from multiple locations and observations, when in principle only two observations were needed. Methods for combining redundant observations were only in their infancy, and would not come to fruition until the work of Legendre, Gauss, and Laplace in the early nineteenth century led to the method of least squares.

Thus contemporary analyses of the 1760s data yielded a variety of answers. A typical case was the analysis of Lalande in 1771, who found values of the earth's mean distance from the sun (the astronomical unit) in the range of ninety-four to ninety-six million miles. But more than a century later in 1891, when locations had been much better determined and mathematical methods improved, Simon Newcomb, dean of late-nineteenth-century American astronomy, determined from the same data a value of 93.0 ± 0.6 million miles, and when he combined the 1761 and 1769 transits with those of 1874 and 1882, he found an overall transit value of 92.95 ± 0.19 million miles.

Before we compare this to a recent determination, let it be said we now know that of the methods developed from the last transit of Venus up to the mid–twentieth century (which included trigonometric parallaxes of asteroids, gravitational perturbations by the sun, and improvements in the constant of stellar aberration), none would surpass in accuracy (although often in claimed precision) the results of the transits of Venus.

Modern astronomy has turned back to Venus to calibrate the astronomical unit, but now in quite a different way. Today giant radio telescopes are used as radar guns, pumping out a tremendously powerful radio signal directed at Venus, and minutes later, switched to receiver mode, detecting the faint echo returning from the planet, the round-trip time being measured by atomic clocks. This interval, together with the speed of electromagnetic waves, yields the distance of Venus at that moment, and thus, through Kepler's third law (see chapter 19), the value of the astronomical unit. The current value stands at 92,955,806.88 ± 0.02 miles. This astonishing result, if taken at its claimed precision, almost defies comprehension. It is the equivalent of measuring the distance between a point in Los Angeles and one in New York with an uncertainty of only a fraction of an inch!

So when the next transit of Venus finally comes along on 8 June 2004, we are not likely to expect new exactitude in determining the astronomical unit, but we might give thought to the words of William Harkness, a key American observer of the nineteenth-century transits, writing just after those transits:

> There will be no other [transit of Venus] till the twenty-first century of our era has dawned upon the earth, and the June flowers are blooming in 2004. When the last [eighteenth-century] transit occurred the intellectual world was awakening from the slumber of ages, and that wondrous scientific activity which has led to our present advanced knowledge was just beginning. What will be the state of science when the next transit season arrives God only knows.

24

The American Kepler

More than a century after his death, the name of Daniel Kirkwood continues to be familiar to almost any student who takes an elementary astronomy course. His discovery of gaps in the distribution of asteroids with distance from the sun made him justly famous, and it is still a staple of elementary texts and research papers alike. But ask any student—or instructor—what else Kirkwood was famous for in his own time, and you will likely be met with a puzzled frown and shake of the head. Tell them that long before the "gap" discovery, he made another that brought him worldwide attention and had prominent astronomers and the popular press hailing him as the equal of Johannes Kepler, and you will likely receive a skeptical smile. But it was so.

Adding to the remarkableness of this event was Daniel Kirkwood's background. He had been born in 1814 on a Maryland farm, and his only education as a child came from the local country school. At the age of nineteen, having no taste for farming, he became a teacher in another such country school. One of his students wished to study algebra, and since Kirkwood knew no algebra, he and the student sat down and worked through an elementary textbook on the subject together—a rather unlikely beginning for one who later held the chair of mathematics at a major university. From this Kirkwood realized that he had both a flair and a taste for mathematics, and so went back to school himself for several years to study the subject. By 1849, he had worked his way up from being a mathematics instructor to becoming principal of an institution called the Pottsville Academy of Pennsylvania. It was about this time that he made his discovery.

During these years of development and reading, Kirkwood had

First published in *American Scientist*, 87:398, September–October 1999.

slowly become intrigued by the fact that while there is a law governing the revolution of planets (Kepler's third law), no law had as yet
been discovered governing their rotations. (Astronomy makes a distinction between the terms *rotate* and *revolve*; a body rotates about
an axis within itself, but revolves about an axis that is exterior. Thus
the earth rotates once a day, but revolves about the sun once a year.)
Kepler's third law of planetary motion says that if P is a planet's period of revolution about the sun and d is its distance from the sun,
then P squared is proportional to d cubed. Kirkwood spent some ten
years mulling off and on over what might be a corresponding law
governing planetary rotations, but with no success.

Eventually, in August 1846, a study of Laplace's theory for the origin of the solar system led him by a process that is not at all clear to
the following proposition. Consider three consecutive planets lined
up in a row. There will be a point between the middle and outer
planet at which a particle will experience equal gravitational force
from the two planets, and another such point between the middle
and inner planet. Calculate the distance, D, between these two points,
the "sphere of influence" of the middle planet. Next, from the known
periods of rotation and revolution calculate the number of rotations,
n, that a planet makes in the course of one revolution. Kirkwood's
calculations suggested to him that n squared was proportional to D
cubed as one went from planet to planet. He referred to his result as
the analog of Kepler's third law.

Unlike some who think they have discovered an important scientific law (Kepler himself, for instance, breaking into paeans of ecstasy
as to how God had waited six thousand years for someone to discover
what Kepler thought to be celestial harmonies among the planets),
Kirkwood behaved in exemplary fashion. He wrote in very modest
fashion to Edward Herrick at Yale, describing his discovery but noting that "perhaps it may be regarded by those better qualified to judge
than myself, as a vagary not worthy of consideration." Herrick suggested that he send his letter to an astronomer at the U.S. Coast Survey, Sears Walker, then well known for his work on Neptune's orbit.
Walker discussed the finding with other members of the American
Philosophical Society, and soon became an enthusiastic advocate,
announcing that it "deserves to rank at least with Kepler's harmonies." In August 1849, Walker presented Kirkwood's letters to a meeting

of the American Association for the Advancement of Science, again concluding the result to be "the most important harmony in the solar system discovered since the time of Kepler, which, in after times, may place their names, side by side, in honorable association."

The AAAS members were impressed. Benjamin Peirce, doyen of astronomy at Harvard, declared it to be "the only discovery of the kind since Kepler's time, that approached near to the character of his three physical laws." Benjamin Gould, founder of what became one of the world's most important astronomy journals, said, "I do not wish to express myself strongly . . . [but] nor can we consider it as very derogatory to the former to speak hereafter of Kepler and Kirkwood together as the discoverers of great planetary harmonies." Newspapers and journals soon brought the public to know of this wonderful and amazing discovery.

Walker, it would seem, was carried away by his own enthusiasm. He submitted a letter to the editor of the prestigious German journal *Astronomische Nachrichten*, outlining "the discovery [as] thus announced by Mr. Kirkwood." It is only one and a half pages long, and contains only one table. The final column is labeled "Kirkwood's diameter of the Sphere of attraction, D," derived, one would assume, from the observational data listed in previous columns of each planet's distance from the sun, mass, and rotational period. If one uses these data to compute the period of revolution by Kepler's third law and thus n for each planet, and plots that against D, one obtains the graph shown in figure 24.1. The straight line has the equation $n = 1000°D^{1.5}$, and is an amazing fit of the line to the points—especially when, as we now know, some of the input data were wildly wrong. The mass of Mercury, for instance, was wrong by a factor of more than two, Venus's rotation period was in error by a factor of more than two hundred, and most of the other data were somewhat in error. What was going on?

The fact is that the statistics were really pretty sparse. Because Mercury has no planet between it and the sun, and Neptune in Kirkwood's time had no known planet beyond it, neither could have a D value calculated for it (although their masses entered into the D values for Venus and Uranus respectively). Moreover, since there is no major planet between Mars and Jupiter, although the asteroids are there and the curious Titius-Bode relation predicts a planet there,

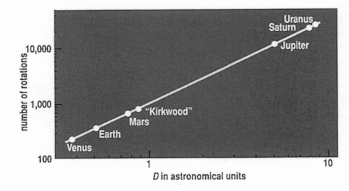

Figure 24.1 *The* n-D *relation as given by the numbers in Walker's letter to the* Astronomische Nachrichten. *The line is described by the equation* n = 1000°D$^{1.5}$. *An astronomical unit is the average distance of the earth from the sun. (Original diagram by Linda Huff,* American Scientist.)

one cannot determine D for either Mars or Jupiter. That leaves Venus, earth, Saturn, and Uranus. The rotational period of Venus (and thus n) was essentially unknown at the time, but vague markings seen telescopically suggested a period near one day, and, Venus being consistently referred to as earth's "sister" planet, that was the value used by Walker and probably Kirkwood. (It is, in fact, 244.3 days, and backward to boot!)

With so little to go on, Walker adopted a two-stage procedure. He ingenuously explained how he first used the data for Venus, earth, and Saturn to determine a preliminary n-D relationship, then used that to work backward to establish that the supposed planet (named "Kirkwood" by Walker) that was thought to have broken up to produce the asteroids must have been 2.908511 times the earth's distance from the sun and had a mass of 0.262290 earth masses and a rotational period of 2.23904 days. With that established, he could then calculate D for Mars, Kirkwood, and Jupiter, and include them in the n-D relationship. Furthermore, this procedure also resulted in Walker's "slight modification" of three other masses, and so, in the end, the marvelous results seen in figure 24.1. Surprise! surprise! No wonder that those like Gould and Peirce, who perhaps never delved into the details, were extreme in their praise of these findings.

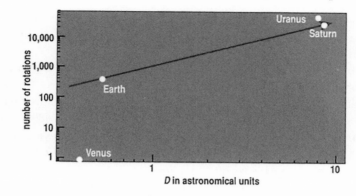

Figure 24.2 *Using modern data and only those planets for which* D *can be calculated* ab initio, *the* n-D *relation is revised. (Original diagram by Linda Huff, American Scientist.)*

Figure 24.2 shows what one gets using today's input numbers and restricting oneself to just Venus, earth, Saturn, and Uranus. Not nearly as exciting! As for the putative planet Kirkwood, it is now virtually certain that no such planet ever existed, the powerful gravity of proto-Jupiter having probably prevented smaller bodies in that region from coalescing into a full-blown planet in the early stages of the solar system, resulting in only a belt of asteroids.

Across the Atlantic, with the element of national pride removed, there was less enthusiasm than in the United States. On the one hand, David Brewster in his 1850 presidential address to the British Association for the Advancement of Science applauded Kirkwood's discovery as a work of genius, but on the other hand, Heinrich Schumacher, editor of the *Astronomische Nachrichten* where Walker had published his letter, declined comment on the subject beyond a scathing note following the letter commenting that readers could judge for themselves "how much . . . is really factual and how much is simply hypothetical." Most European astronomers seem to have sided with Schumacher, if indeed they commented on the matter at all.

Eventually, however, all the high praise died down, and, it seems, less and less was heard about Kirkwood's analogy, until today it is all but completely forgotten. Still, Kirkwood received an honorary doctorate from the University of Pennsylvania in 1852, and appointment to the chair of mathematics at Indiana University in 1856, events that one presumes would probably not have happened without all the previous praise and publicity.

For one so modest and diffident, it is remarkable that the setback in fortune of the n-D relation did nothing to dissuade Kirkwood from publishing later research. Three books and well over a hundred papers came later, the most famous, in 1866, examining the statistics of the eighty or ninety asteroids then known and finding that essentially none of them were located where their periods of revolution about the sun would have been simple fractions (one-half, one-third, three-fifths, etc.) of Jupiter's period. These were the now famous gaps in the asteroid belt. Kirkwood recognized that asteroids in such resonances encountered the full effect of Jupiter's pull more frequently than otherwise, and so were soon removed from those positions. In fact, he realized that more sophisticated effects were present as well, and indeed, research on the subject has continued into our own times. Kirkwood also pointed out the same year that the main gap in Saturn's rings (Cassini's division) is due to the same phenomenon: ring particles at that location would have one-third the period of the satellite Enceladus and nearly one-half that of Mimas. In 1861, he gave the first convincing evidence of a connection between meteor showers and comets, and over the years made several other solid contributions to solar-system science.

So in the end Kirkwood could look back on a very satisfactory career, even if he wasn't the American Kepler. He retired to California in 1886, becoming for a while an astronomy lecturer, aged seventy-seven, at Stanford University, and eventually died in 1895, a few months short of his eighty-first birthday. His friend and former colleague, Joseph Swain, said of him at the time of his retirement "that during his fifty years as a teacher, he has gained from his students such universal love and admiration as few men can enjoy, and while . . . he has made many valuable contributions to science, as a genial, temperate, and genuine man, he has solved the problem of gracefully growing old." Kepler should have been so lucky.

Eclipse Vicissitudes:
Thomas Edison and the Chickens

On 11 August 1999, my wife and I found ourselves in the countryside of southern Hungary, preparing to watch a total eclipse of the sun. Luck was with us; an overcast, rainy morning gave way to clear skies an hour or two before totality began near midday. Although it wasn't our first total solar eclipse, we were nevertheless caught up in the fascination the event always brings: the increasingly eerie quality of the dying light as the moon relentlessly gnaws its way across the face of the sun, the expectant hush of the staring people around you, the sudden cries of awe as, in this case, the diamond-ring phenomenon suddenly flashes into view. And with the closing darkness the local birds, true to form, set off to roost in the nearby trees, twittering in bewilderment at the early coming of night. I was immediately reminded of the story of Thomas Edison and the chickens.

It's a story that centers on the total eclipse of 29 July 1878, an eclipse notable on more than one historical front (see chapter 14), and one well above average in interest to astronomers. The path of total eclipse, starting in Alaska, was to sweep down the spine of the Rocky Mountains and out across the Gulf of Mexico before ending in Cuba. This meant that it might be observed from an area with generally clear weather containing many high-altitude mountain sites, mostly above the haze and water vapor that plague lower sites. This fact took on added importance in that the later nineteenth century was a time in which astrophysics, the study of stars themselves as distinct from their positions and motions, was beginning an explosive growth.

So it was that Samuel Langley, director of the Allegheny Observatory, intended to use these auspicious circumstances to investigate

First published in *American Scientist*, 88:120, March–April 2000.

the strange outer atmosphere of the sun called the corona. Until recently this phenomenon, then visible only at a total solar eclipse, had been thought by some to be the moon's atmosphere rendered visible by backlighting from the sun. Langley wanted to find out more about the corona, in particular its temperature. But how to measure the temperature of something ninety-three million miles away? The obvious thought was that if the corona is hot, as one would expect if it is solar, it must produce infrared radiation, albeit difficult to measure at that distance. This was a real problem for Langley, since existing infrared detectors (thermopiles) were generally too insensitive to yield a useful result. And it was here that Thomas Edison entered the story.

Edison, although only thirty-one at the time, was already the most famous inventor in the world, having the previous year invented the phonograph, something that amazed people everywhere. Working at all hours, he directed a twelve-man laboratory at Menlo Park, New Jersey, turning out astonishing inventions almost day by day. He already held eighty-nine patents in telegraphy alone, had invented the stock ticker and the carbon-button telephone, and was just that year announcing the incandescent light bulb, although the latter was still a year from completion. This was typical of Edison, who never hesitated for a moment when it came to making seemingly wild and certainly premature announcements. All in all he was a forceful and brash young man, and the popular press loved him for it. To judge from the space accorded him in newspapers of the day, he must have had a reporter at his side almost constantly.

By early 1878 he had, in fact, already been experimenting with a new invention that was particularly sensitive to heat, and had discussed it briefly in correspondence with Langley. So it was to Edison that Langley turned in his quest for an instrument that could measure the corona's heat during the forthcoming eclipse. He noted that the necessary instrument would have to be at least one hundred times more sensitive than existing thermopiles. Edison saw no difficulty in that. Langley's best thermopile was capable of detecting a change of about 0.0001 degree Fahrenheit in temperature, and Edison's new invention was already about eight times more sensitive.

Edison called his invention a tasimeter, a name that took him more time and worry to invent than the instrument itself. Its basis was the carbon button, already invented as a transducer for telephones. The infrared radiation was focused onto a vulcanite rod. Its heat caused the rod to expand and press against the button of powdered graphite. Since the electrical resistivity of powdered carbon is extraordinarily sensitive to pressure, the output of the instrument was read as a deflection of a galvanometer incorporated in an electrical circuit with the button. The thing certainly worked. Naturally there were demonstrations for the press, and Edison liked to show how easily the tasimeter could detect the heat from a person's hand thirty feet away, while one correspondent found it to be so sensitive "that let a person come into the room with a lighted cigar, and it will drive the little animal wild." Edison's favorite demonstration was to show that the tasimeter was six times more sensitive to heat from his little finger than was a thermopile to a red-hot iron. After some further tweaking, Edison claimed the tasimeter had a sensitivity of a millionth degree Fahrenheit, and so met Langley's challenge. There is some doubt, however, as to whether Edison ever really made a sensitivity test in these terms.

But time was moving on, and Langley needed to make some astronomical tests before setting off to the eclipse. In early June, he wrote Edison asking that the latter send a tasimeter to the Allegheny Observatory for tests, as well as additional carbon buttons for tests with other instruments. The carbon buttons arrived, but no tasimeter. Well into June, Langley sent a reminder of the "promised tasimeter which I shall have great pleasure in testing." Then "I expect to go in the beginning of July to observe the solar eclipse," and finally on 5 July, just days before leaving, a terse telegram: "Send by express to Allegheny. I leave Monday." There was no reply from Edison.

What Langley didn't know was that Edison himself was going to the eclipse, armed with the tasimeter. Henry Draper, a wealthy medical doctor with an interest in astronomy, had invited Edison to join his party at the eclipse site in Rawlins, Wyoming Territory. Edison, in need of a break from Menlo Park, treated the trip as a vacation and gladly accepted.

He would travel free as a courtesy of the Union Pacific Railroad Company, which further provided him with a letter introducing him as "Mr. Edison, the celebrated inventor and telegrapher," and instructing telegraphers along the route to "send all messages of Mr. Thomas A. Edison free." The New York press turned out in full to see him off at the Pennsylvania Railroad depot and record his parting words: "Yes," said Thomas Edison, "it [the tasimeter] will measure any degree of heat that can be measured. If the sun's corona has any heat of its own . . . the tasimeter will measure it accurately." That evening the *Daily Graphic* devoted its full front page to Edison and the tasimeter. En route to the Rockies, Edison was accompanied by a correspondent of the *New York Herald* and hailed at every stop by the local press and railroad telegraphers, who regarded him as one of their own. Edison, one presumes, was never at a loss for a reply.

Rawlins, population eight hundred, had been chosen as an eclipse site by some scientists since it was nearly seven thousand feet in altitude and, in particular, was readily accessible by train. Others, however, were spread up and down the mountain ranges, including Langley, 250 miles southward down the eclipse path on fourteen-thousand-foot Pikes Peak, still hoping to measure the coronal heat with a thermopile. All told, the astronomers observing the eclipse included some of the most illustrious names from around the world. They were doubtless unamused to find themselves summed up by the press in the phrase "Professor Edison, accompanied by a party of scientists," especially since Edison was young, boastful, knew little astronomy, and was not a professor to boot. Their feelings can be judged from the five-hundred-page report on the eclipse eventually prepared by the U.S. Naval Observatory: the name of Edison appears nowhere in it!

Edison was more explicit regarding his contempt for most academics. He is on record with the statement "I wouldn't give a penny for the ordinary college graduate, except those from Institutes of Technology . . . they aren't filled up with Latin, philosophy, and all that ninny stuff." As for the mathematical sciences, "I can hire mathematicians at $15 a week but they can't hire me." He was scornfully amused by the precise latitude and longitude determinations made at the eclipse site: "It seemed to take an immense amount of mathematics. I preserved one of the sheets which looked like the timetable of a Chinese railroad."

One writer has described Edison at Rawlins as unwittingly like "a rather typical modern-day eclipse-goer since (1) he made preparations only shortly before leaving, (2) he elected to defer final assembly and tests until arrival at the site, (3) he claimed success immediately after third contact, (4) he never reduced his data, and (5) he never published his scientific findings." Which brings us finally to the chickens. This oft-told tale has been recorded by J. A. Eddy thus:

> When Edison stepped off the train at Rawlins he found the professional astronomers already ensconced in the best rooms of the only hotel and already in possessive claim of the more protected places from which to observe the coming eclipse. All that remained for the tasimeter was a dilapidated hen-house, and in its doorway Edison set up his telescope and equipment. In the afternoon of 29 July, as totality neared, a brisk Wyoming wind arose, filling the darkening sky with dirt and debris. These conditions made the balancing of the tasimeter . . . especially difficult, and with the onset of darkness at second contact, the tasimeter was still not adjusted. Only two minutes of totality remained. Feverishly he worked, but alas! With the sun covered and sky dark, the chickens came home to roost, through Edison's observatory door, past the telescope, in, around, and over the frantic inventor. Uninitiated in astronomy, he had failed to allow for a fundamental eclipse phenomenon.

What degree of credence one can place in this story is highly uncertain; one suspects a good deal of gleeful embellishment with the passing years. Edison himself, dictating his memoirs some thirty years later, tells of setting up his equipment "in a small yard enclosed by a board fence six feet high; at one end there was a house for hens. I noticed they all went to roost just before totality. At the same time a slight wind arose and at the moment of totality the atmosphere was filled with thistle down and other light articles." This was the account that made the official biographies, of course, but the truth probably lies somewhere in between.

What is certain, however, is that Edison did in the end get a shot at the corona with his tasimeter. The *New York Herald* reporter, in his dispatch home, noted, "When but one minute of totality remained Edison succeeded in crowding the light from the corona upon the small opening of the tasimeter. Instantly the galvanometer cleared its boundaries. Edison was overjoyed." Edison himself, however, in later years faced up to the irony of having made too much of an improvement over existing detectors: "My apparatus was entirely too sensitive and I got no results." Langley, by contrast, found himself once again the victim of an inadequate thermopile, which failed to detect the corona.

As for the tasimeter itself, it faded into obscurity rather rapidly. For a while there was talk of commercial applications (iceberg detector on ships, navigator's sun finder in cloudy weather, etc.), but it proved to be a slow, nonlinear, poorly repeating, and highly unstable device. In short, great for qualitative demonstrations but useless for quantitative measurements. Edison got in a parting shot by dedicating it without patent to "the dilettantes in the higher branches of science."

Langley had his revenge by inventing the bolometer a couple of years later. Much better than a thermopile, it was less sensitive than the tasimeter, but it was stable, gave readings repeatable to 1 percent, and could be used over a wider range of the spectrum. In announcing it Langley made no comparison with or even mention of the tasimeter.

History perhaps had the last laugh. It was later pointed out that during a total solar eclipse thirty-six years before the 1878 one, Professor Luigi Magrini, observing at Milan with an unusually sensitive thermopile on a reflecting telescope, had already measured a definite infrared signal from the solar corona. It had long been overlooked. Not, I suppose, that Edison would have cared much. Ninny stuff.

In Search of Better Skies:
Harvard in Peru. I

Like many an old codger before me, I'm rather given these days to reflecting on the changes that have transpired over the years of my career. Naturally, I quite often come to the standard old-codger conclusion that things have pretty much gone to the dogs, but one thing that is now indisputably better is the life of the observational astronomer. When I was a graduate student in the mid-fifties, one almost always had to carry out observations with whatever local telescope was available in whatever climate one's location offered, usually far from the best. National or consortium facilities at carefully chosen first-rate sites were mostly yet to come. When they did, they resulted in a generation of astronomers in what seemed almost perpetual flight between home and observing sites. Today we are returning to a world of stay-at-homes, but now sitting in front of their computers, using the Internet to control a telescope and data acquisition half a world away, if not in outer space.

Nevertheless, the idea of creating field stations at desirable locations, where data could be obtained and sent back to home base for analysis, goes back a long way. It received particular impetus in the later nineteenth century, when photography entered astronomy. Photographs could be taken relatively quickly at the distant site, and then returned home for leisurely analysis by experts. Take the case of the Harvard College Observatory.

It began with the death of Uriah Boyden in 1879, at the age of seventy-four. Boyden had been a practical Boston engineer-inventor of some repute, and probably surprised everyone when he left nearly a quarter of a million dollars to whatever astronomical institution could

First published in *American Scientist*, 88:396, September–October 2000.

convince his trustees that it would build an astronomical observatory on a mountain "at such an elevation as to be free, so far as practicable, from the impediments to accurate observations which occur in the observatories now existing, owing to atmospheric influences." Boyden had been mildly eccentric, but his heirs saw this as eccentricity run rampant, especially since it comprised almost his entire fortune. A legal battle of considerable proportions ensued, but eventually the will was declared valid. Among those who took note was Edward Pickering, director of the Harvard College Observatory, who had for some time had an interest in the field-station idea. In fact, he wanted two: one in the Northern Hemisphere and one in the Southern, so the entire sky would be available for Harvard's research. There were further years of delay while the trustees consulted endlessly, but eventually, in 1887, Pickering won the Boyden Fund for Harvard.

Pickering decided on southern California for the northern station, and word soon reached him that there was already local enthusiasm for establishing an observatory outside Pasadena on what was then called Wilson's Peak. The local enthusiasm came about because northern Californians had recently acquired the Lick Observatory a little south of San Francisco. (Lick had been a millionaire who had been persuaded, with some difficulty, to build the observatory as a monument to himself instead of a pyramid rivaling that of Cheops in downtown San Francisco. He is buried in the concrete pier of one of the observatory's telescopes.) The Lick Observatory, run by the University of California at San Francisco, boasted a 36-inch refracting telescope—the world's largest. The Pasadena enthusiasts aimed for a 40-inch refractor on Wilson's Peak, to be run by the University of Southern California.

Pickering wasn't keen on a big refractor; photography was becoming the dominant tool of observational astronomy, and it was better served by reflecting telescopes. However, it was agreed that Harvard would send an expedition to test the Wilson's Peak site, and share the test results with the public group. The test showed this peak to be "an astronomer's paradise," as one observer put it, and it seemed that with good will and cooperation, both groups could establish themselves there. But in what seems to have been an absolutely farcical mistake, a Harvard lawyer drew up papers inadvertently referring to Harvard's cooperation with the University of California instead of the University of Southern California, and this reached the Pasadena group before Pickering caught the error. The group was outraged: Harvard was selling them out to their northern rivals! There had been some friction before this, and now the pot boiled over. One might think it would have been simple enough to correct this error, but as it turned out the two sides were never reconciled, and in the end Harvard abandoned its Californian ambitions.

There remained an irony or two, however. The Pasadena group proceeded with its 40-inch refractor plans and found enough funding to have such a lens actually cast before the funding collapsed and those plans also had to be abandoned. Later, a Harvard-trained astronomer, George Ellery Hale, persuaded a wealthy businessman, Charles Yerkes, to buy the unused lens for a new observatory established in Yerkes's name and directed by Hale for the University of Chicago. Subsequently Hale, under the aegis of the Carnegie Institution of Washington, established an observatory on Wilson's Peak, now known as Mount Wilson, which after 1918 included the 100-inch telescope described in chapter 1, with which the likes of Edwin Hubble and Walter Baade revolutionized modern astronomy. And later, as we saw in chapter 16, a Mount Wilson astronomer, Harlow Shapley, became director of the Harvard College Observatory. So in the end, the southern Californians got the world's largest telescope after all, while Harvard found satisfaction in its indirect contributions and connections thereto.

Harvard's proposed Southern Hemisphere station got off to a less controversial start. Given the comparative slowness of travel and communications in the late nineteenth century, Pickering had opted for a site as close to home as possible, and since much of the Pacific coast of South America lies almost due south of Boston, this was the area investigated. Existing weather records indicated that the town of Chosica, some thirty miles inland from Lima in Peru, would be a good starting point.

Solon Bailey was the man chosen to lead the expedition. He was then thirty-three and known for his determination and resourcefulness, having started at the Harvard College Observatory as an unpaid assistant working forty hours a week. His proven abilities soon led to not only a paying position there, but an M.A. degree as well. Now, in early 1889, with his wife and young son and his brother, Marshall, he left for two years in South America to choose a site for and establish a southern field station.

Marshall had the unenviable task of shepherding some one hundred crates of the expedition's materiel through the journey south, including a crossing of Panama by train. They passed the partly dug canal, on which the De Lesseps Company had recently stopped work, including "the company's hospital and burial ground. The vastness of each bore witness to the power of the forces with which promoters of the canal were obliged to contend. . . . Decay and desolation" were everywhere, and "a depression will doubtless follow as soon as this enterprise shall be definitely abandoned."

After nearly two months of travel, the party arrived at Chosica and immediately realized that establishing an observatory there was out of the question. The little town, in the verdant valley of the Rimac River, was pleasant, but nearby "abrupt, precipitous mountains" shut out much of the sky. A day or two later, Solon and Marshall began scouting the neighboring terrain for a better site. Led by local guides, they started off in a northerly direction, passing several well-preserved but long abandoned settlements. Solon later remarked that "I should place the population of the valley near Chosica in the days of the Incas at six thousand. Today there are perhaps five hundred. This . . . well illustrates how Peru has changed since she fell into the hands of the Spanish conquerors." The going was extremely arduous; at times "shut in by steep walls that rose a thousand or more feet on either side, the breeze faded away, but the sun beat down with great heat. We followed along the dry bed of what must at some time have been a torrent of tremendous power; for it had worn a rut fifty feet deep into the bed of the ravine, and had brought down great numbers of bowlders, some of enormous size." Later they were climbing hand and foot up a slope of several thousand feet or more, until in early afternoon they were atop a mountain they thought would do for their field station. They dubbed it Mt. Harvard. "In every direction nothing but barren mountains were to be seen. . . . To the north and south we looked down into gloomy ravines thousands of feet deep." Returning to the hotel in early evening, "We looked back at the dark outline [of Mount Harvard] against the sky; the thought of a residence on that isolated spot brought a strange sense of gloom and loneliness."

Clearly, they would need much local help in setting up their observatory on Mount Harvard (or any other mountain in the region), and in any case, they had some misgivings about that site and thought they should seek further weather advice before making a definite decision. So it was that they now returned to Lima for a couple of weeks; first for an audience with the Peruvian president to ask for manual help, and second to consult whatever weather experts they might find in the capital.

The president proved stern but cordial ("with battle-scarred countenance . . . his face showed marks of privation and passion"), and issued orders "compelling the governor of any province to furnish peons to assist in transporting our instruments." The weather experts, on the other hand, gave much conflicting advice, although the town of Arequipa was mentioned favorably several times, and "Mr Ralph Abercrombie, the English meteorologist . . . spoke very highly of the Atacama desert [in northern Chile]."

Solon decided to do an initial exploration farther up the Rimac valley, and after returning to Chosica briefly, the brothers set off for Chicla, some thirty miles east. This was no great distance, but they were now climbing ever higher into the Andes and enduring tropical heat. "The road . . . is simply a mule path four feet wide, trailing along the face of the mountain, in many places lofty, steep, and slippery. Frequently it winds around the face of a cliff, cut into the solid rock, and the path worn smooth by innumerable feet for hundreds of years has nothing whatever to prevent the careless from slipping over the side and falling hundreds of feet into the river below. To add to the difficulty, one meets in a day hundreds of laden mules, donkeys, and llamas. The only safety is to insist upon the inside track. Accidents are not rare."

They passed the Verrugas River, a small tributary of the Rimac. "From this river . . . is named the peculiar and dangerous disease known as Verrugas. From it arose a great mortality among [local] workmen. It is still common. . . . The disease is characterized by intense pain, but, especially, by the appearance on different parts of the body of sacs filled with blood. These are sometimes of considerable size, and when the number of them is great, the loss of blood is considerable. Nevertheless, it is regarded as a favorable symptom to have the disease appear on the outside, and medicines are taken to produce this result. It was thought to be caused by drinking Verrugas water."

The following day they breakfasted in San Mateo at an altitude of ten thousand feet, and here Marshall succumbed to his first bout of mountain sickness. "It manifested itself by dizziness, faintness, and nausea; complete unconsciousness occurred twice for a few minutes duration. The patient was placed on the ground, and bruised garlic, the odor of which is thought by the natives to have great efficacy, was provided in abundance. A little hot soup, however . . . speedily brought the patient into better condition." Solon's report remarks on the unpredictability of mountain sickness: later in their travels Marshall had a very serious attack at only seven thousand feet, but three days later at nearly fifteen thousand feet he experienced only mild dizziness. Solon himself could go to about sixteen thousand feet without discomfort, but above that he suffered "violent nausea and vomiting, which continued for several hours until I descended to a lower altitude."

Although this short trip provided "the wildest . . . scenery it has been my privilege to see . . . in any country," it revealed no site well suited to an astronomical observatory, and since Pickering was doubtless anxious for some useful work to be done, Solon decided they would set up their equipment on barren Mt. Harvard after all, even though "no supplies were at hand, and all food and water would need to be carried several miles" daily. Mentally, he made a note to himself that during the mild cloudy season he had heard about, he and Marshall would explore much farther afield, checking Arequipa and the Atacama Desert, among other sites.

The story was far from over. Mount Harvard had to be abandoned eventually, Pickering made a disastrous mistake that almost aborted the entire project, and Solon and his family found themselves in the midst of a revolution, taken hostage at one point and later preparing to defend their home with revolvers.

In Search of Better Skies:
Harvard in Peru. II

The year is 1889. Thirty-four-year-old Solon Bailey, his brother Marshall, wife Ruth, and young son Irving are in Peru to establish an astronomical field station for Harvard University. In the last chapter we saw that they had tentatively chosen for their station a mountain some thirty miles inland from Lima, while the two brothers explored a little farther inland for a possibly better site. Finding nothing better, they rejoined Ruth and Irving and set to work taking astronomical photographs and measuring the magnitudes of southern stars from Mt. Harvard.

It was tough going. The mountain was desolate indeed; it "furnished neither water nor food. These were brought daily from Chosica eight miles away and nearly four thousand feet below. Considerable difficulty was experienced in finding a trustworthy man to act as muleteer, [and] . . . we were repeatedly disappointed in the non-arrival of our day's water and provisions." Snakes were frequently encountered, as were scorpions ("one was met with in a shoe, another in a coat sleeve"), and seven-inch tarantulas. "Our life was so isolated that man and animal, dog, cat, and goat were on terms of the greatest intimacy and equality." Nevertheless, the two men and their Peruvian assistant, Elias Vieyra, made good progress with their astronomical work. At least they did at first, but with the coming of spring and early summer later in 1889 there came also persistent high, thin cloud, which was enough to halt their delicate observations.

Solon finally realized that Mt. Harvard would not do. He and Marshall must travel much farther afield to find a better site. They had already heard, as noted in chapter 26, that Arequipa, five hundred miles to the southeast, was a gorgeously sunny location, and

First published in *American Scientist*, 89:123, March–April 2001.

that the Atacama Desert, another six hundred miles to the south in Chile, was even better. So, leaving Elias to do what he could on Mt. Harvard and installing Mrs. Bailey and Irving in comfortable quarters in Lima, Solon and Marshall set off on a four-month tour of promising sites.

They sailed from Lima on 14 November and arrived in Mollendo, the port for Arequipa, on the seventeenth. Ships had to anchor a mile offshore, and the swell was often so bad that passengers were taken in a cage by barge to shore, where a crane lofted them to dry land. It was a completely barren region of sand and rock; the water supply was piped in from a hundred miles away, and "all food of every kind, except such fish as are caught in the vicinity, must be brought by rail or boat."

Inland, the Baileys traveled by train where possible, otherwise by stage coach, mule, or—on occasion—by foot. It was by train that they approached Arequipa, some sixty miles from Mollendo. And what a difference! Arequipa, located in the valley of the Chili River, greeted them with "fields of waving grain and groves of fruit-trees. . . . The first view of the city is really beautiful, surpassing in picturesqueness any other Peruvian city we had seen. Above the city, which rests just at its foot, rises . . . El Misti, a nearly extinct volcano about nineteen thousand feet high."

Although clearly impressed, the brothers spent only a day there before leaving for La Paz in Bolivia. Again they went by train to Puno on the shores of Lake Titicaca, climbing ever higher into the Andes. The train traveled a considerable distance at an altitude near sixteen thousand feet, and "of those present in the car somewhat more than half were quite ill from mountain sickness." It was early summer, but bitterly cold in Puno, yet "while I shivered in my overcoat, Indian women with bare feet and legs and open breasts seemed happy and comfortable." They left Puno at five in the morning on a small steamer to cross Lake Titicaca, passing islands where they "were able to see quite distinctly the ruins of the ancient buildings known as temples and palaces," including "the so-called Temple of the Virgins of the Sun." La Paz, however, was a disappointment: "'After Paris, La Paz,' say the Bolivians, and we decided that it was a long way after."

By late November the brothers were back in Arequipa, where Solon spent a week in bed recovering from some malady picked up en route. This unexpected delay provided further evidence of the clear and pleasant climate of that location.

Soon, however, the two were aboard ship again and sailing as far south as Valparaíso in Chile. Christmas Day was spent in Santiago, but there was little there to satisfy their site-seeking, and by 1 January 1890 they were at Antofagasta in northern Chile, preparing to examine the potential of the Atacama Desert. It is one of the driest and most desolate places on earth, and "utterly barren. . . . Not even the cactus gains the least foothold here." The only practical way to proceed was to take the railroad inland away from the coastal fogs, to where a nitrate-mining company operated. Even this could be a trifle uncertain, since "work is carried on wherever the yields are the richest, and when expedient the whole town, buildings, and population [of a thousand men] are shifted many miles along the railway." Water came from deep wells and was brought in some distance by donkey. "[It] has an unpleasant taste, but is said to be wholesome. As I could find no one who drank water, however, I do not know on what authority the statement was based." Nevertheless, the Baileys spent almost a month at this spot, Pampas Central, and were amazed at the clarity of the sky. Twenty-eight of their twenty-nine nights there were cloudless, and the transparency such that Solon could clearly see eleven stars of the Pleiades with his naked eye. The brothers' photometric measures of stars were precise and repeatable.

On 5 March they arrived back at Mt. Harvard with Ruth and Irving. Almost no work had been accomplished in their absence, the weather having been terrible. In fact, torrential rains had severely damaged their living quarters and almost washed away the rain gauge itself. Solon wasted no time in writing home to Edward Pickering, advocating that Mt. Harvard be abandoned in favor of either Arequipa or a site in the Atacama Desert. He noted that the latter offered the best sites, but that running a station in so utterly desolate a place would be extremely difficult and impose great hardship on its staff. (If Solon could have seen a century into the future, he surely would have been pleased to find that later technology overcame most of the difficulties, so that some of the largest and most important astronomical observatories are now located in the high Atacama Desert.)

Pickering soon made up his mind. Instructions were issued that all equipment on Mt. Harvard be packed and removed to Arequipa, which would be home to Harvard's permanent southern field station. He went further. The Baileys had endured nearly two harsh years finding a site, and once it was established at Arequipa they should return home to New England. Who then would run the new station? Here Edward Pickering made what was possibly the worst decision of his life. He would send his own brother, William.

Meanwhile, though, the Bailey team undertook the move to Arequipa. The hardest part was just getting everything off Mt. Harvard and down to Chosica, there being no road, and the materiel including their living quarters. "Several mules, made unsteady by loads of lumber, rolled down the mountain-side for some distance. No bones were broken, however, and no special damage done. All the instruments were either carried by hand or on the backs of mules that were led by hand."

By the end of the year, the Baileys had rented a suitable house a mile or two outside Arequipa and begun astronomical observations from its roof. Choice of a permanent location awaited the arrival of William Pickering, his family, and his assistants, which took place in January 1891. William quickly decided on a nearby site for a permanent observatory, and some time was spent moving everything there and setting up the instruments. Finally, on 15 May, "we [the Baileys] bade farewell to Arequipa, and two days later we sailed south. We had decided to return to the United States by the Straits of Magellan and Europe. The journey along the west coast was relieved from monotony by incidents in the Chilian civil war." They arrived home on 15 August.

William Pickering had formerly been a physics instructor at MIT, but had joined the Harvard Observatory in 1887 at the invitation of his brother, to help run the new department that involved the field-station projects. In that regard it wasn't surprising that he should be sent to Peru; what was surprising was that Edward failed to foresee the ensuing uproar. It was not long in coming. William had been told that initially he was not to spend more than five hundred dollars leasing accommodations for the new station, in case the site was not as good as expected. But the first communiqué Edward had from William was a four-word cable stating simply, "Send four thousand more." In reply to Edward's demand for explanation, William announced that he had bought a considerable tract of land outright and was preparing to build a substantial house for himself, his wife, two children, mother-in-law, nurse, and several assistants, for which he would need another seven thousand dollars, plus two thousand more for running expenses. Edward must have come close to a heart attack. The Boyden Fund, which was paying for all this, could not possibly sustain such a rate of expenditure. He was on the verge of closing the entire project and demanding William's return, but after he had consulted with Harvard's president, William's requests were reluctantly met. A severe admonishment against wild expenditures was sent, along with Edward's expressed hope that scientific results would soon be forthcoming.

Edward had laid out a detailed plan for photographic sky surveys and was most anxious for results to justify the excessive expenditures. "I would give up everything to keep the telescopes running all night, the plates developed, and sent on promptly," he told William. But nine months later, not one plate had arrived in Cambridge. A curt cable ordering William to "photograph with thirteen inch [telescope]" produced no results.

The fact was that William was happily engaged on something entirely different. Some years earlier, in 1877, an Italian astronomer, Giovanni Schiaparelli, had noted markings on Mars which he called 'canali,' meaning 'channels.' Mistranslated as 'canals,' the word started a popular theory of alien beings that lasted many years. William was now busy with his eye to the telescope to see what he could do with Mars. He saw no reason to restrict his announcement of results to the scientific literature, let alone Edward, and instead cabled them to the *New York Herald*. So it was that Edward's first intimation of what his brother was actually doing was a major newspaper report that William had discovered great mountain ranges on Mars, that the polar icecaps of Mars were melting to form rivers flowing toward the Martian equator, and that he had seen at least forty lakes ranging up to a hundred miles in size. The astronomers of the Lick Observatory in California were reported to have received this intelligence "with a kind of amazement." Working under equally good conditions with a telescope three times the size of William's, they were quite unable to see any of these things.

The long-suffering Edward again chastised his brother. "The telegram to the N.Y. Herald has given you a colossal newspaper reputation. A flood of cuttings have appeared, forty-nine coming this morning. . . . You would have rendered yourself less liable to criticism if you had stated your interpretations were probable instead of . . . certain." William was undaunted. Jupiter was next on his list, and soon Edward was reading in the *Herald* that his brother had determined that "the first satellite [of Jupiter] is egg-shaped and revolves end over end. . . . Its period is twelve hours and fifty-five minutes." The Lick astronomers hardly knew how to express their views politely. "Very likely the telegrams are wrong?" their director inquired hopefully of Edward, but of course they were not.

Finally (one might say at long last) Edward took action. William must return to Cambridge, and Bailey would replace him in Arequipa. William, of course, was outraged and became truculent. "I've accomplished a pretty big thing . . . and have got the [Harvard] authorities a great deal for their money," he wrote. "This is *my* Observatory . . . and it is not Bailey's Observatory nor anybody else's," and so forth. But Edward was implacable. William tried pleading, promising to get the photographic work done. Edward was unmoved.

On 25 February 1893, Solon, his wife, and son arrived back in Arequipa, and after what must have been an extraordinarily unpleasant month of overlap with William, set to work cleaning up and starting the set program of observing. A revolution in Peru was looming, however, and the Baileys would find themselves embroiled. We'll see in the concluding chapter what happened.

In Search of Better Skies:
Harvard in Peru. III

One must wonder what Solon Bailey's feelings were when, on 25 February 1893, he found himself once again in Arequipa. William Pickering had completely wasted the field station's resources and brought great disrepute to Harvard's work there. Now Bailey had to resume direction, William having been—with considerable difficulty—recalled to Cambridge.

It was a daunting task. Bailey found the station close to ruin, both financially and in its equipment. Assistants and servants had been given carte blanche to run up bills in the town, while William and his family had not stinted themselves either. Telescopes were grossly maladjusted, having been modified by William for purposes for which they were not designed. The seismograph recorded a major earthquake every time the door to its room was opened, some weather instruments had never been set up, and the all-important clocks were seriously unreliable, some with parts missing.

Bailey was equal to the task. Under his wise and firm leadership, the station slowly regained its stability and its good working order. The golden years of astronomy in Arequipa finally began. The main work was to photograph the southern skies (repeatedly in interesting areas) with various instruments and thereby to derive the positions, magnitudes, and spectra of stars and other objects. It might have seemed dull, routine work to many, but it led to some of the most profound developments in astronomy during the twentieth century. Photographs from Arequipa, for example, led to the discovery of the Cepheid period-luminosity law, a cornerstone of modern cosmology.

One thing Bailey decided against, however, was observing Mars at

First published in *American Scientist*, 89:402, September–October 2001.

another of its nearer passages. On the previous such occasion, as we have seen, William Pickering, using the field station's equipment, had claimed in widespread newspaper reports the most absurd results about lakes and canals on Mars. Writing with some irony to Edward Pickering, Bailey reported, "I have given very brief attention to Mars. I have concluded that my time will be better spent in other lines, especially as the northern observatories will doubtless give so much attention to it and see all there is to be seen, if not more, and also I fear that I have not the creative faculty sufficiently developed to make a mark as an observer of Mars."

Instead, Bailey turned his attention to meteorology. Arequipa, at an elevation less than eight thousand feet, barely qualified as the sort of mountainous observatory site stipulated by the supporting grant from the Boyden Fund, especially with the Andes towering nearby. Was there a better site at a higher elevation not far off? Bailey soon found the question could not be immediately answered, because Peru then lacked almost any systematic meteorological data. He therefore determined to set up a roughly east-west array of weather stations, running from Mollendo on the Pacific coast right across to the eastern side of the Andes, some in valleys, others on mountains.

More easily said than done! Bailey noted in one of his reports: "In Peru, outside of one or two large towns, there are no highways suitable for vehicles of any kind. Travel and traffic, except that done by the few railways, are carried on entirely on the backs of horses, mules, donkeys, and llamas." But characteristically there was no question of Bailey staying in Arequipa while others did the hard work. Although now in his fortieth year, Bailey was soon out on horseback getting things done. As his obituarist would eventually note, "There were adventures such as swinging from peak to peak over a deep valley in a cage suspended by a single cable, or descending rapidly from elevations of 10,000 feet in a hand car with gravitation as the motive power, or passing nights on desolate mountains where perhaps the only sounds were the 'ripple of the [river] far below, and the flapping of the condor's wings.'"

Bailey's greatest interest, though, was much nearer Arequipa itself. The most notable sight for a newcomer was beautiful snow-capped El Misti, the dormant nineteen-thousand-foot volcano only twelve miles beyond the town. Bailey must have realized that putting a manned observatory at that altitude was impractical, but the weather data would be useful, and there were always the lower slopes.

Just climbing El Misti was a major undertaking. It had been climbed at least as early as 1784, when a party of priests had succeeded in erecting an iron cross on the peak, but others had died in the effort. Conditions could be extreme. One party of climbers, Bailey reported, "arrived at the summit during a frightful tempest, with terrific thunder and lightning. . . . The bodies of those present were electrified, so that when they lifted their hands, they could hear the discharge of the electricity from the tips of their fingers." Another pair had climbed at temperatures near minus eleven degrees Fahrenheit, their hands lacerated and bloody from the ice and lava rock.

Bailey himself had set out to climb El Misti during his earlier stay in Arequipa, but mountain sickness had left him unconscious just below the sixteen-thousand-foot level, and it had been necessary for his companions to carry him down. He wrote that "the forms in which this malady manifests itself are oppression of the lungs and difficulty in breathing, more or less violent headache, nausea and vomiting, dizziness and faintness, sometimes reaching entire unconsciousness, nervousness, sometimes tending to delirium, and rarely hemorrhage from the nose, eyes, and ears." Pulse rates, he found, typically went from the normal seventy per minute to well over a hundred, even after a good rest.

However, Bailey had also made two significant observations: first that mountain sickness became much more pronounced when the victim was nearing exhaustion from the climb itself, and second that mules seemed less affected by altitude than humans. He thus concluded that if a rough path could be cleared to some point well up on the mountain, it might be possible to ride mules to that point and then set off in fresh condition to tackle the remaining portion on foot. The local people scoffed at the idea, but Bailey had such a path constructed and found he could take mules to nearly the eighteen-thousand-foot level—the highest mules had ever been taken, so far as he knew. Even so, the remaining thousand or more feet were not easy:

> Panting for breath, stopping to rest at every three or four steps, often struggling on hands and knees, we kept on, hardly believing there could come an end, when, suddenly, we were there. There was no introduction; we did not come to the crater; the crater came to us. The whole

view was spread out before us in an instant as if a curtain had been drawn. All things conspired to produce surroundings which few have seen and none described. The great altitude, the enormous craters, the sulphurous vapors, the drifting clouds, the deep shadows cast by the setting sun, the inexplicable but deep depression of spirits caused by exhaustion and illness, combined to produce the profoundest sense of awe.

It had taken the expedition eight days to get there.

So Bailey in due course had his meteorological station atop El Misti—the highest such station in the world, he noted with satisfaction. Even so, the weather records had to be collected and the instruments reset every week or so, and it turned out there was only one person on Bailey's staff who could do this regularly without suffering the dreaded mountain sickness. Still, writing to Pickering in October 1893, Bailey expressed satisfaction and looked forward "to the time when scientists will regularly ascend to great heights for meteorological study, by means of captive balloons or flying machines."

It was about this time that new problems began to appear. There were increasing rumors in Arequipa that Peru was in for a revolution. At first Bailey was quite jocular about it, writing to Pickering that he might "have to remove the lenses and use the telescope tubes for cannon." Later, his tone became more serious. "We have two or three revolvers and with the addition of a few good clubs, I think we should be able to keep off any drunken rabble." He went on to say that they were laying in provisions to withstand a siege, ordering heavy wooden shutters for their windows and doors, preparing to bury the telescope lenses in a deep hole beneath the floorboards of the house (while hoping any necessity for that would happen in the cloudy season), and so forth. Pickering responded enthusiastically, advising them to pour buckets of hot water from the upstairs windows on any persons attempting to force an entrance.

There was rioting in Arequipa in late 1894, but the first real taste of trouble came in January 1895, when Bailey, his wife, and young son were traveling to Mollendo by train. "We heard a tremendous shout of 'Viva Pierola.' I looked out only to see a crowd of men armed with rifles and revolvers come rushing around the train and into the car. The car was at once filled with cries of 'Jesus Maria' and 'Por Dios.'" However, the revolutionaries behaved "with great moderation," and apart from locking the passengers in the train, "offered us no indignity whatever." At Mollendo the passengers remained locked in while the rebels captured the town, an act that was accomplished almost immediately, there being only fifteen soldiers guarding it. The Baileys spent the night in the house of the shipping agent there, Bailey and their host standing guard all night with club and revolver respectively.

Soon after the Baileys returned to Arequipa, the town was besieged and the telegraph line cut, and savage fighting broke out. The Americans put their emergency plans into operation, and after burying the telescope lenses (it *was* the cloudy season), ran the American flag up over the house and settled down to wait things out. Each night, though, one of them would creep out in the darkness to retrieve the day's weather records, even though there was rifle fire only fifty feet away on occasion. They were never attacked.

When it was all over and the revolutionaries had won, Bailey decided it would be prudent to establish friendly relations with the new president-to-be by inviting him for a tour and reception at the observatory. "The expense was moderate, about twenty dollars," he reported to Pickering, "and as Pierola is sure to be the next president, if he lives, I think it was a wise act."

And so the Arequipa station settled down to long years of steady work. Bailey himself developed an estimable scientific reputation for his work on variable stars in globular clusters, and today his name is best remembered for the class of such stars named after him. His travels were not over, though. In 1908, again at Pickering's request, he traveled through much of South Africa in search of a site possibly better than Arequipa. This at a time when that country had not yet been fully established as a self-governing British dominion, and was still recovering from the Anglo-Boer War of a few years earlier. Bailey did choose a site, but it was not until 1927 that Harvard decided to close its Arequipa station and move the operation to South Africa.

It is not entirely clear what factors entered into the decision to close the station in Peru. That it was cloudier than initially thought seems to have been the principal reason. There is a story, though, told me by a colleague familiar with Arequipa, that in the later years observing was largely in the hands of an employee who developed a drinking problem and thereafter would all too often simply enter "cloudy" in the observing log before heading for the downtown bars. I wouldn't bet on it though.

Edward Pickering died in 1919, and Bailey became acting director of the observatory at Harvard for two years, before Harlow Shapley was appointed. Shapley had the highest regard for Bailey, both for the accuracy of his work and because it was he who first suggested that Shapley study variable stars, through which Shapley arrived at his epochal studies of the size of our galaxy and our position in it.

William Pickering outlived his brother by nearly twenty years, spending the later part of his life running an observatory in Jamaica, from which emanated a stream of reports on canals, snow-capped mountains, and vegetation on the moon. As he proudly proclaimed, "I have seen everything practically except the selenites [inhabitants of the moon] themselves running around with spades to turn off the water into other channels!"

Solon and Ruth Bailey loved their years in Arequipa. One colleague recalled that "many Arequipa friends as well as American visitors enjoyed the informal teas on the Observatory balcony over the garden, constantly full of luxuriant flowers, with the beautiful view of the rushing Chile River, up to majestic El Misti, framed in the tropical sky."

Solon died in 1931, at the age of seventy-six, at his home in Norwell, Massachusetts. One obituary notes that "he won the respect of all by his wide sympathy, his justice, his never-failing kindness, and his complete lack of self-seeking." One is reminded of a remark by Edward Pickering many years earlier that "science is an ennobling pursuit only when it is unselfish." No less, he might have added, is that true of life itself.

Notes

The Ghost of Mount Wilson

It is pleasing to report that the 100-inch telescope and other facilities of the Mount Wilson Observatory continue to flourish, although now under the aegis of the Mount Wilson Institute. This nonprofit, tax-exempt organization was formed in 1986. Its purpose is to realize for public benefit the maximum scientific and educational potential of the Mount Wilson site and facilities. Descriptions of the work done there now, along with many pictures, new and old, may be found at *http://www.mtwilson.edu/*.

The advent of the Hubble Space Telescope in 1990 prompted a number of Hubble biographies and the publication of some previously unpublished papers of Hubble's. Some of these suggest that a number of the earlier stories about the man may be apocryphal. As Donald Osterbrock, Ronald Brashear, and Joel Gwinn (in'"Evolution of the Universe of Galaxies," ed. Richard G. Kron, *Astronomical Society of the Pacific Conference Series*, 10:1, 1990) note, "Hubble's early life . . . has become a myth. He has been the subject . . . of supposedly factual biographies which bear little relation to his own real life."

Candid Posterity and the Englishman. II

Halley's tombstone was later removed from his grave and is now seen by innumerable visitors in the outside wall of the old Royal Greenwich Observatory.

Bloody Sirius

At least two important papers have appeared on this topic since this article was written. (See the references for complete bibliographical information.) Douglas C. B. Whittet shows that the only plausible physical reason that Sirius could have been seen to be red in Ptolemy's time was that it was observed close to the horizon. Similarly, Roger C. Ceragioli gives strong nonastronomical reasons that the star was referred to as being red. The mystery is therefore now much diminished, if not completely solved.

Stonehenge and the Archaeoastronomers

As noted at the end of the chapter, not much has been written about the astronomy of Stonehenge since the 1960s, but astronomers and archaeologists have developed much more respect for one another. In particular, astronomers have realized that one cannot blindly produce archaeoastronomical theories without considering the archaeological record as well. In general, while archaeoastronomy (and its close relative, ethnoastronomy) has spread to include sites worldwide, the idea that ancient stone monuments were built specifically as astronomical observatories has lost favor.

The Tunguska Event

A much greater interest has developed in recent years regarding possible collisions between the earth and asteroids, comets, and large meteorites, and there are now several observatories devoted to searching for such objects that might be on a collision course with us. One website that provides updates on this is *http://impact.arc.nasa.gov*. No candidates have been detected so far.

Comets—Again!

In the event, the impact of the cometary fragments on Jupiter was not directly seen in visible light. Nor were there any visible reflections seen from Jovian satellites, and distant spacecraft recorded only modest flashes. However, telescopes working at infrared wavelengths did record spectacular effects, and long after the event, dark clouds at the impact points in Jupiter's upper atmosphere were readily visible even in small telescopes.

The Great Debate

See the note to chapter 1.

The Mysterious Gamma-Ray Bursters

The development of the rapid-response system to marshal optical, radio, and x-ray detectors immediately following detection of a gamma-ray burst paid off within six months of this article being written, allowing gamma-ray bursters to be studied at other wavelengths. The emerging picture is that the large majority of them are indeed at cosmological distances, while a few are more local. This suggests there is more than one kind of burster. Favored theories invoke the collapse of an exceptionally massive star, or the collision and merging of two neutron stars initially orbiting one another. Details, however, remain sketchy.

References

Candid Posterity and the Englishman

Armitage, Angus. *Edmond Halley*. London: Thomas Nelson & Sons, 1966.

MacPike, Eugene Fairfield. *Correspondence and Papers of Edmond Halley*. Oxford, U.K.: Oxford University Press, 1932.

Ronan, Colin Alistair. *Edmond Halley—Genius in Eclipse*. New York: Doubleday, 1969.

Bloody Sirius

Ceragioli, Roger C. "Behind the 'Red Sirius' Myth." *Sky & Telescope*, 83:613, 1992.

Whittet, Douglas C. B. "A Physical Interpretation of the 'Red Sirius' Anomaly." *Monthly Notices of the Royal Astronomical Society*, 310:355, 1999.

Stonehenge and the Archaeoastronomers

Hawkins, Gerald. *Stonehenge Decoded*. Garden City, N.Y.: Doubleday, 1965.

Hoyle, Fred. *From Stonehenge to Modern Cosmology*. San Francisco: W. H. Freeman, 1970.

Ruggles, Clive. *Astronomy in Prehistoric Britain and Ireland*. New Haven, Conn.: Yale University Press, 1999.

Alexander Thom and Archaeoastronomy

Heggie, Douglas (ed.). *Archaeoastronomy in the Old World*. Cambridge, U.K.: Cambridge University Press, 1982.

Ruggles, Clive (ed.). *Records in Stone: Papers in Memory of Alexander Thom*. Cambridge, U.K.: Cambridge University Press, 1988.

Ruggles, Clive. *Astronomy in Prehistoric Britain and Ireland*. New Haven, Conn.: Yale University Press, 1999.

Thom, Archibald Stevenson. *Walking in All of the Squares: A Biography of Alexander Thom*. Argyll, Scotland: Argyll Publishing, 1995.

Wood, John Edwin. *Sun, Moon, and Standing Stones*. Oxford, U.K.: Oxford University Press, 1978.

The Shape of the Earth

Smith, James Raymond. *From Plane to Spheroid*. Rancho Cordova, Calif.: Landmark Enterprises, 1986.
von Hagen, Victor Wolfgang. *South America Called Them*. New York: Alfred A. Knopf, 1945.

The Last Universalist

Buttmann, Günther. *The Shadow of the Telescope: A Biography of John Herschel*, trans. Bernard Pagel. New York: Charles Scribner's Sons, 1970.
Evans, David S., Terence J. Deeming, Betty H. Evans, and Stephen Goldfarb. *Herschel at the Cape: Diaries and Correspondence of Sir John Herschel 1834–1838*. Austin: University of Texas Press, 1969.

The Great Moon Hoax

A full reprinting of the Great Moon Hoax appeared in *The Sky* (now *Sky & Telescope*), vol. 1, nos. 4, 5, 6, 1937.

The Tunguska Event

Baxter, John, and Thomas Atkins. *The Fire Came By: The Riddle of the Great Siberian Explosion*. New York: Doubleday, 1976.
Hills, Jack G., and M. Patrick Goda. "The Fragmentation of Small Asteroids in the Atmosphere." *The Astronomical Journal*, 105:1114, 1993.
Sekanina, Zdenek. "The Tunguska Event: No Cometary Signature in Evidence." *The Astronomical Journal*, 88:1382, 1983.

Comets—Again!

Sekanina, Zdenek. "Disintegration Phenomena Expected during Collision of Comet Shoemaker-Levy 9 with Jupiter." *Science*, 262:382, 1993.

In Pursuit of Vulcan

Eddy, John A. "The Great Eclipse of 1878." *Sky & Telescope*, 45:340, 1973.
Narlikar, Jayant V., and N. C. Rana. "Newtonian N-body Calculations of the Advance of Mercury's Perihelion." *Monthly Notices of the Royal Astronomical Society*, 213:657, 1985.
Proctor, Richard A. *Old and New Astronomy*. London: Longmans, Green & Co., 1882.

The Neptune Affair

Grosser, Morton. *The Discovery of Neptune*. Cambridge, Mass.: Harvard University Press, 1962.

Kowal, Charles T., and Stillman Drake. "Galileo Saw Neptune." *Nature*, 287:311, 1980.

The Great Debate

Berendzen, Richard, Richard Hart, and Daniel Seeley. *Man Discovers the Galaxies*. New York: Neale Watson Academic Publications, 1976.

Hoskin, Michael A. "The 'Great Debate': What Really Happened." *Journal for the History of Astronomy*, 7:169. 1976.

The Extraordinary and Short-Lived Career of Jeremiah Horrocks

Applebaum, Wilbur. "Jeremiah Horrocks." *Dictionary of Scientific Biography*, ed. Charles C. Gillispie, 6:514, 1972.

Chapman, Allan. "Jeremiah Horrocks, the Transit of Venus, and the 'New Astronomy' in Early Seventeenth-Century England." *Quarterly Journal of the Royal Astronomical Society*, 31:333, 1990.

Weber, R. L. *A Random Walk in Science*. New York: Crane, Russak & Co., 1973.

The Mysterious Gamma-Ray Bursters

Nemiroff, Robert J. "The 75th Anniversary Astronomical Debate on the Distance Scale to Gamma-Ray Bursts." *Publications of the Astronomical Society of the Pacific*, 107:1131, 1995.

Transits, Travels, and Tribulations

Dick, Steven J., Wayne Orchiston, and Tom Love. "Simon Newcomb, William Harkness and the Nineteenth-Century American Transit of Venus Expeditions." *Journal for the History of Astronomy*, 29:221, 1998.

Hogg, Helen Sawyer. "Out of Old Books: Wales's Journal of a Voyage in 1768." *Journal of the Royal Astronomical Society of Canada*, 42:153, 189, 1948.

Hogg, Helen Sawyer. "Out of Old Books: Le Gentil and the Transits of Venus, 1761 and 1769." *Journal of the Royal Astronomical Society of Canada*, 45:37, 89, 127, 173, 1951.

Proctor, Richard A. *Transits of Venus*. London: Longmans, Green & Co., 1882.

Villiers, Alan John. *Captain James Cook*. New York: Charles Scribner's Sons, 1967.

Woolf, Harry. *The Transits of Venus: A Study in Eighteenth-Century Science*. Princeton, N.J.: Princeton University Press, 1959.

The American Kepler

Numbers, Ronald L. "The American Kepler: Daniel Kirkwood and His Analogy." *Journal for the History of Astronomy*, 4:13, 1973.

Walker, Sears C. "Letter." *Astronomische Nachrichten*, 30:11 (no. 697), 1850.

Eclipse Vicissitudes: Thomas Edison and the Chickens

Eddy, John A. "Thomas A. Edison and Infra-red Astronomy." *Journal for the History of Astronomy*, 3:165, 1972.

Eddy, John A. "The Great Eclipse of 1878." *Sky & Telescope*, 45:340, 1973.

In Search of Better Skies: Harvard in Peru

Bailey, Solon I. "History of the Expedition." *Harvard College Observatory Annals*, 34:1, 1895.

Bailey, Solon I. "Peruvian Meteorology 1888–1890." *Harvard College Observatory Annals*, 39:1, 1899.

Jones, Bessie Zaban, and Lyle Gifford Boyd. *The Harvard College Observatory: The First Four Directorships, 1839–1919*. Cambridge, Mass.: Harvard University Press, 1971.

Index

About the Author

Donald Fernie was born in South Africa, but has lived in Canada for the past forty years. He currently is professor emeritus in the Department of Astronomy and Astrophysics at the University of Toronto.